SEMICONDUCTING POLYMERS

POLYMERS

Synthesis and Photophysical Properties

SEMICONDUCTING POLYMERS

Synthesis and Photophysical Properties

Edited by
Raquel Aparecida Domingues, PhD
Daniel Henrique do Amaral Corrêa, PhD

First edition published 2022

Apple Academic Press Inc.
1265 Goldenrod Circle, NE,
Palm Bay, FL 32905 USA
4164 Lakeshore Road, Burlington,
ON, L7L 1A4 Canada

CRC Press
6000 Broken Sound Parkway NW,
Suite 300, Boca Raton, FL 33487-2742 USA
2 Park Square, Milton Park,
Abingdon, Oxon, OX14 4RN UK

© 2022 Apple Academic Press, Inc.

Apple Academic Press exclusively co-publishes with CRC Press, an imprint of Taylor & Francis Group, LLC

Library and Archives Canada Cataloguing in Publication

Title: Semiconducting polymers : synthesis and photophysical properties / edited by Raquel Aparecida Domingues, PhD, Daniel Henrique do Amaral Corrêa, PhD.
Names: Domingues, Raquel Aparecida, editor. | Corrêa, Daniel Henrique do Amaral, editor.
Description: Includes bibliographical references and index.
Identifiers: Canadiana (print) 20200308084 | Canadiana (ebook) 20200308130 | ISBN 9781771888684 (hardcover) | ISBN 9780367816186 (ebook)
Subjects: LCSH: Conducting polymers. | LCSH: Polymers—Electric properties. | LCSH: Semiconductors.
Classification: LCC QD382.C66 S46 2021 | DDC 547/.70457—dc23

Library of Congress Cataloging-in-Publication Data

Names: Domingues, Raquel Aparecida, editor. | Corrêa, Daniel Henrique do Amaral, editor.
Title: Semiconducting polymers : synthesis and photophysical properties / edited by Raquel Aparecida Domingues, PhD, Daniel Henrique do Amaral Corrêa, PhD.
Description: First edition. | Palm Bay, FL : Apple Academic Press, 2021. | Includes bibliographical references and index. | Summary: "Semiconducting polymers are of great interest for applications in electroluminescent devices, solar cells, batteries, and diodes. This volume, Semiconducting Polymers: Synthesis and Photophysical Properties, provides a thorough introduction to the basic concepts of the photophysics of semiconducting polymers, as well as a description of the principal polymerization methods for luminescent polymers. Divided into two main sections, the book first introduces the advances made in polymer synthesis, and then goes on to focus on the photophysics aspects, also exploring how new advances in the area of controlled syntheses of semiconducting polymers are applied. An understanding of the photophysics process in this kind of material requires some knowledge of many different terms in this field, so a chapter on the basic aspects is included. The process that occurs in semiconducting polymers spans time scales that are unimaginably fast, sometimes less than a picosecond. To appreciate this extraordinary scale, it is necessary to learn a range of vocabularies and concepts that stretch from the basic concepts of photophysics to modern applications, such as electroluminescent devices, solar cells, batteries, and diodes. This book provides a starting point for a broadly based understanding of photophysics concepts applied in understanding semiconducting polymers, incorporating critical ideas from across the scientific spectrum. The book is intended for graduate students in a range of disciplines. It will also prove valuable for faculty and instructors, and scientists and researchers in chemistry, physics, and engineering"-- Provided by publisher.
Identifiers: LCCN 2020035791 (print) | LCCN 2020035792 (ebook) | ISBN 9781771888684 (hardcover) | ISBN 9780367816186 (ebook)
Subjects: LCSH: Polymers. | Semiconductors. | Polymerization.
Classification: LCC QD382.S4 S475 2021 (print) | LCC QD382.S4 (ebook) | DDC 620.1/9204297--dc23
LC record available at https://lccn.loc.gov/2020035791
LC ebook record available at https://lccn.loc.gov/2020035792

ISBN: 978-1-77188-868-4 (hbk)
ISBN: 978-1-77463-792-0 (pbk)
ISBN: 978-0-36781-618-6 (ebk)

About the Editors

Raquel Aparecida Domingues, PhD
Professor, Federal University of São Paulo (UNIFESP), São José dos Campos, SP, Brazil

Raquel Aparecida Domingues, PhD, is Professor at the Federal University of São Paulo (UNIFESP), São José dos Campos, SP, Brazil. Her research focuses on the photophysics of fluorescent siloxanes and semiconducting polymers. She obtained her PhD at the University of Campinas (UNICAMP) in 2013. She is the author of several papers published in various journals and conference proceedings.

Daniel Henrique do Amaral Corrêa, PhD
Professor, São Leopoldo Mandic College of Medicine, Araras, SP, Brazil, and União Mogiana para o Desenvolvimento da Educação (UNIMOGI), Mogi Guaçu, SP, Brazil

Daniel Henrique do Amaral Corrêa, PhD, is Professor at the São Leopoldo Mandic College of Medicine, Araras, SP, Brazil, and União Mogiana para o Desenvolvimento da Educação (UNIMOGI), Mogi Guaçu, SP, Brazil. He obtained his PhD at the University of Campinas (UNICAMP) in 2010. He is the author of several papers in the field of protein biochemistry and several abstracts in conference proceedings.

Contents

Contents

Contributors

Luís Gustavo Teixeira Alves Duarte
Chemistry Institute, University of Campinas, Campinas, Brazil

Diego De Azevedo
Chemistry Institute, University of Campinas-UNICAMP, P.O. Box – 6154, CEP 13083-970, Campinas, SP, Brazil

Cristina A. Barboza
Institute of Physics, Polish Academy of Sciences, 02 668, Warsaw, Poland

Natália Ferreira Braga
Institute of Science and Technology of the Federal University of São Paulo (ICT-UNIFESP), 12231-280, São José dos Campos, SP, Brazil

Fernando Henrique Cristovan
Institute of Science and Technology of the Federal University of São Paulo (ICT-UNIFESP), 12231-280, São José dos Campos, SP, Brazil

Raquel Aparecida Domingues
Institute of Science and Technology, Federal University of São Paulo-UNIFESP, R. Talim, 330, 12231-280, São José dos Campos, SP, Brazil

Flavio S. Freitas
The Federal Institute of Education, Science and Technology of the South of Minas Gerais, Pouso Alegre Campus, Maria da Conceição Santos Avenue 900, 37560-260, Pouso Alegre, MG, Brazil

Jilian Nei De Freitas
CTI – Renato Archer Information Technology Center, Dom Pedro I Highway, km 143.6, 13069-901, Campinas Brazil

José Carlos Germino
Chemistry Institute, University of Campinas, Campinas, Brazil

Rossano Lang
Institute of Science and Technology of the Federal University of São Paulo (ICT-UNIFESP), 12231-280, São José dos Campos, SP, Brazil

Rodrigo A. Mendes
Biophotonics Laboratory, CePOF-IFSC/USP, São Carlos, Brazil

Andreia De Morais
CTI – Renato Archer Information Technology Center, Dom Pedro I Highway, km 143.6, 13069-901, Campinas Brazil

Rafael Felipe Coelho Neves
The Federal Institute of Education, Science and Technology of the South of Minas Gerais, Poços de Caldas Campus, Dirce Pereira Rosa Avenue 300, 37713-100, Poços de Caldas, MG, Brazil

Erick Piovesan
Institute of Physics, Federal University of Uberlândia-UFU, 38400-902, Uberlândia, MG, Brazil

Roberson Saraiva Polli
Institute of Science and Technology, Federal University of São Paulo, 12231-280,
São José dos Campos, SP, Brazil

Mariana Silva Recco
Institute of Science and Technology of the Federal University of São Paulo (ICT-UNIFESP),
12231-280, São José dos Campos, SP, Brazil

Débora Aparecida Ribeiro
Institute of Science and Technology of the Federal University of São Paulo (ICT-UNIFESP),
12231-280, São José dos Campos, SP, Brazil

Gustavo Targino Valente
São Carlos Institute of Physics, University of São Paulo, PO Box – 369, 13560-970, São Carlos,
SP, Brazil, E-mail: gtvfisica@gmail.com

Nirton Cristi Silva Vieira
Institute of Science and Technology, Federal University of São Paulo, 12231-280,
São José dos Campos, SP, Brazil

Abbreviations

ADF	Amsterdam density functional
APD	avalanche photodiode
BBL	poly(benzimidazobenzophenanthroline ladder)
BDMO-PPV	poly[2,5-bis(3,'7'-dimethyl-octyloxy)1,4-phenylene-vinylene]
BDT	benzodithiophene
BT	benzothiadiazole
CASPT2	complete active space second-order perturbation theory
CCD	charge-coupled diode
CCL	color-conversion layer
CCSD	coupled-cluster singles and doubles
CLSM	confocal laser scanning microscopy
CT	charge-transfer
CTS	charge transfer state
D-A	donor-acceptor
DBS	dodecylbenzene sulfonic
DFT	density functional theory
DOS	density of state
DPBT	4,7-diphenyl benzothiadiazole
DPDSB	2,5-diphenyl-1,4-distyrylbenzene
DTBT	4,7-di-thienyl benzothiadiazole
EL	electroluminescence
ERB	erythrosin B
ETL	electron-transporting layer
FCS	fluorescence correlation spectroscopy
FCT	field configurable transistor
FET	field-effect transistor
f-PVK	fully overlapped
FRET	Förster resonance energy transfer
FWHM	full width at half maximum
GGAs	generalized gradient approximations
GRIM	Grignard's metathesis
H_2O	water

HF	Hartree-Fock
H-H	head-to-head
HIL	hole injection layer
HK	Hohenberg-Kohn
HOMO	highest occupied molecular orbital
H-T	head-to-tail
IC	internal conversion
ISC	intersystem crossing
ITO	indium tin oxide
KS	Kohn-Sham
LDA	local density approximation
LEGSs	local exciton ground states
LGS	local ground state
LUMO	lowest unoccupied molecular orbital
MDMO-PPV	poly[2-methoxy-5-(3', 7'-dimethyloctyloxy)-1,4-phenylene vinylene]
MEH-PPV	poly(2-methoxy-5-(2-ethylhexyloxy)-1,4-phenylenevinylene)
MM	molecular mechanics
MP2	Møller-Plesset
Ni(dppe)Cl$_2$	[1,2-bis(diphenylphosphino)ethane]nickel(II) dichloride
Ni(dppf)Cl$_2$	[1,1'-bis(diphenylphosphino)ferrocene]nickel(II) dichloride
Ni(dppp)Cl$_2$	[1,3-bis(diphenylphosphino)propane]nickel(II) dichloride
Ni(PPh$_3$)$_2$Cl$_2$	bis(triphenylphosphine)nickel(II) dichloride
NPs	nanoparticles
OC1C10-PPV	poly(2-methoxy-5-(2',6'-dimethyloctyloxy)-p-phenylenevinylene)
ODOS	occupied density of states
OLED	organic light-emitting diode
OLET	organic light-emitting transistor
P3HOT	poly(3-hexyloxythiphene)
P3HT	poly(3-hexylthiophene)
P3PhT	poly(3-phenylthiophene)
PA	photoinduced absorption
PAN	polyacrylonitrile
PC	polycarbonate

PCzSF	poly[2′,7′-bis(3,6-dioctylcarbazo-9-yl)-spirobifluorene]
PDPB	poly(diphenyl)butadiene
PDPV	poly-4,4'-diphenylene diphenylvinylene
PE	polyethylene
PEO	poly(ethylene oxide)
PEPPSI-Ipr	[1,3-bis(2,6-diisopropyl phenol)imidazole-2-ylidene] (3-chloropyridyl)palladium(II) dichloride
PET	poly(ethylene terephthalate)
PF6s	poly(9,9-dihexylfluorene)s
PFCB	perfluorocyclobutyl aryl ether
PFO	polyfluorenes
PFOPen	polyfluorene derivative
PFP	poly[(9,9-dioctylfluorenyl-2,7-diyl)-alt-co-(9,9-di-{5'-pentanyl}-fluorenyl-2,7-diyl)]
PFPBr$_2$	poly[9,9-bis-(60-bromohexylfluoren-2,7-diyl)-*alt*-co-(benzen-1,4-diyl)]
PL	photoluminescence
PLA	polylactic acid
PLEDs	polymeric light-emitting diodes
PLQE	photoluminescence quantum efficiency
PMMA	poly(methyl-methacrylate)
PPEB	poly(*p*-phenylene-ethynylene-butadiynylene
PPP	poly(para-phenylene)
PPV	poly(p-phenylene vinylene)
PPV-DP	poly-(1,3-phenylene diphenylvinylene)
PPVs	poly(para-phenylene vinylenes)s
PPy	polypyrrole
PS	polystyrene
PSF	polyspirobifluorene
PT	polythiophene
PU	polyurethane
PVA	poly(vinyl alcohol)
PVA	polyvinyl acetate
PVB	poly(vinyl butyral-co-vinyl alcohol-co-vinyl acetate)
PVC	polyvinyl chloride
PVK	partially overlapped
QM	quantum mechanics
RB	Rose Bengal

RGB	red, green, and blue
RI	regioirregular
RR	regioregular
SA	self-assembly
SB	Strickler-Berg
SemPolys	semiconducting polymers
SM	Suzuki-Miyaura
SMFS	single-molecule fluorescence spectroscopy
SPCM	single-photon counting module
SPPO13	9,9'-spirobi-(fluorene)-2,7-diylbis(diphenylphosphine oxide)
SR	singlet radioactive
SS	Stokes Shift
SSH	Su-Schrieffer-Heeger
SVA	solvent vapor annealing
SW	super yellow
T	transmission
TAS	transient absorption spectroscopy
TBP	thiophene-2-boronic acid pinacol ester
TD-DFT	time-dependent density functional theory
THF	tetrahydrofuran
TP	p-terphenyl
TPA	triphenylamine
TPB	tetraphenyl butadiene
TPD	tetraphenyldiaminobiphenyl
TR	triplet radiative
T-T	tail-to-tail
UV/Vis	ultraviolet and visible regions
VR	vibrational relaxation
WOLED	white organic light-emitting diodes
XC	exchange-correlation
Xe	xenon

Symbols

[ ]	molar concentration
°C	degree Celsius
Br	Bromo
c	speed of light
D_a	Dalton
$E_{(gap)}$	band gap energy
E_a	electron affinity
f	oscillator strength
I	ionization potential
I_i	initial intensity of the radiation
k_F	emission rate
k_{nr}	non-radiative deactivation rates
M	transition dipole moment operator
Mn	number average molecular weight
Mw	weight average molecular weight
N	Avogadro constant
n	refractive index
Pd	palladium
S	Huang-Rhys parameter
S_1	first excited electronic state
W	Watt

GREEK SYMBOLS

σ	sample effective irradiated cross-section
τ	emission lifetime
ψ	electronic wavefunctions
χ	vibrational wavefunctions
ε	extinction coefficient
φ	quantum yield
γ	the ratio between the exciton formation within the device and the number of electrons circulating in the external circuit

λ	wavelengths
ν	vibrational level
σ^2	the variance of the density of state

Preface

This book is an introduction to the basic concepts of the photophysics of semiconducting polymers, as well as a description of the principal polymerization methods for luminescent polymers. Thus, the book is divided into two main sections; the first introduces the advances made in polymer synthesis, and the second focuses on the photophysics aspects, including theoretical issues and applications. The understanding of the photophysics process in this kind of material requires some knowledge of many different terms in this field. Thus, he book dedicates one chapter to basics aspects. Also, our discussion of the process that occurs in semiconducting polymers will span time scales unimaginably fast, sometimes less than a picosecond. To appreciate this extraordinary scale, it is necessary to learn a range of vocabularies and concepts that stretch from basic concepts of photophysics to modern applications such as electroluminescent devices, solar cells, batteries, and diodes. This book attempts to provide the starting point for a broadly based understanding of photophysics concepts applied in understanding semiconducting polymers, incorporating critical ideas from across the scientific spectrum. The book is intended for graduate students in a range of disciplines. An understanding of the basic principles of chemistry and physics is assumed.

CHAPTER 1

Polymerization Methods for Luminescent Conjugated Polymers

DÉBORA APARECIDA RIBEIRO[1], NATÁLIA FERREIRA BRAGA[1],
MARIANA SILVA RECCO[1], ERICK PIOVESAN[2], ROSSANO LANG[1] and
FERNANDO HENRIQUE CRISTOVAN[1]

[1]*Instituto de Ciência e Tecnologia – ICT, UNIFESP, 12231-280,
São José dos Campos, SP, Brazil*

[2]*Instituto de Física, Universidade Federal de Uberlândia – UFU,
38400-902, Uberlândia, MG, Brazil*

1.1 INTRODUCTION

There are several comprehensive reviews of different organic synthesis techniques in the literature to produce luminescent conjugated polymers. The purpose here is just to provide a quick reference guide for the reader. Therefore, this chapter presents a brief introduction to the main strategies employed in synthesizing luminescent conjugated polymers.

Over the years, it has become clear that the polymeric structure plays a critical role in determining the physicochemical properties of organic electronic materials. The regularity control and the spatial organization of the polymer chain structure lead to the remarkable and exciting tuning of the optoelectrical properties. In other words, the synthesis conditions directly affect the structural and electronic distribution since the conformation of the primary chain is responsible for the intrinsic physicochemical properties.

Great efforts have been devoted to studying synthesis methods to promote desired changes in the polymeric composition and the development of new luminescent organic materials. It is up to the researcher to understand for then decide the best synthetic route to be applied to his challenge. Frequently, the monomers may not undergo an efficient reaction through the selected method. In this way, the choice of a suitable synthetic approach is crucial to minimize structural defects and maximize the desired properties (polymers with a high molar mass, π molecular orbitals overlap, high luminescence, etc.).

Most advances in electronic plastics technology are primarily due to the improvements and innovations achieved in the synthesis of conjugated polymers (Morin et al., 2016). Conjugated polymers are organic molecules formed by a backbone chain of alternating double-single bonds between carbon atoms (or heteroatom), allowing electrons to flow in specific conditions. They have been considered promising materials for a wide diversity of organic electronic technologies, once they exhibit robust and useful structural arrangements demanded by several applications (Friend et al., 1999; McCullough, 1998; Akcelrud, 2003). The main-chain conjugation associated with a doping process can make the polymers extrinsically conducting or semiconducting (Kanatzidis, 1990; Sheina et al., 2005). From there, the features of metals and inorganic semiconducting have been explored, trying to preserve the properties unique to polymers.

The conjugated conducting polymers were discovered and developed by the Shirakawa group (2001), along with the Heeger (2001) and MacDiarmid and Epstein (1991) researchers. The discovery awarded them the Nobel Prize in Chemistry in the year 2000. The first report of a luminescent organic semiconductor, the anthracene, came out in the 1960s. In 1990, the first conjugated polymer, the poly(p-phenylene vinylene) (PPV), was reported (Friend et al., 1999). The PPV molecule is composed of a benzene ring and vinyl bonds, as shown in Figure 1.1. Since then, the PPV polymer has been extensively studied for light-emitting device applications (Li et al., 1999; Bao et al., 1999).

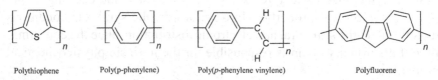

Polythiophene Poly(p-phenylene) Poly(p-phenylene vinylene)   Polyfluorene

FIGURE 1.1 Chemical structure of some luminescent polymers.

It is known that the light-emitting intensity of the luminescent polymers (some shown in Figure 1.1) depends crucially on the structural features of the polymer chain, such as the chain regularity in substituted conjugated polymers (Andersson et al., 1994; Akcelrud, 2003; Shinde et al., 2012). Hence, aiming to increase the regioregularity in conjugated polymers, several polymerization methods were developed, and some will be briefly described as follows.

In general, substituted conjugated polymers can be classified by the organization of the polymer chain, named as regioregular (RR) or regio-irregular (RI). Heterocyclic polymers, such as polythiophenes (PTs), are typically coupled at the 2- and 5-position, with a substitution generally at the 3- or 4-position of the thiophene ring (McCullough and Lowe, 1992; Leclerc and Faid, 1997; Müllen et al., 2014). Figure 1.2 exhibits the possible coupling forms (combinations) for dimers and trimers of substituted PTs. For homocyclic polymers, the presence of substituted side chains may occur in different positions, depending on the material type, and according to the more stable configuration of the polymer chain (Akcelrud, 2003). The substitution of side chains in the primary structure can influence several properties such as redox activity, charge mobility, absorption and optical emission, chemical reactivity, molar mass, and solubility (McCullough and Lowe, 1992; Akcelrud, 2003; Seeboth et al., 2014).

The 2-position is called "head," and the 5-position is called "tail." These are the positions for the possible coupling of the polymer main chain, specifically for heterocyclic chains. The dimers can be connected in several configurations such as head-to-head (H-H), tail-to-tail (T-T), and head-to-tail (H-T). Indeed, as the size of the principal chain increases, the orientation becomes more complex (McCullough, 1998; Müllen et al., 2014). The planarity of the principal polymer chain is established by an effective molecular superposition (a regular molecular structure), which generates narrower bandgaps and lower oxidation potentials (Koeckelberghs et al., 2006). Luminescent polymers that do not possess side-chain substitution have a low solubility in organic solvents (Maior et al., 1990). The low solubility is due to the strong intermolecular attraction of the polymer chain with adjacent chains, causing packaging and aggregations. In that case, it is hard for the solvent to penetrate the polymeric structure (Leclerc and Faid, 1997). For example, unsubstituted PTs can be formed by acid polymerization and present a low solubility in organic solvents (Armour et al., 1967; Müllen et al., 2014). On the other hand, significant

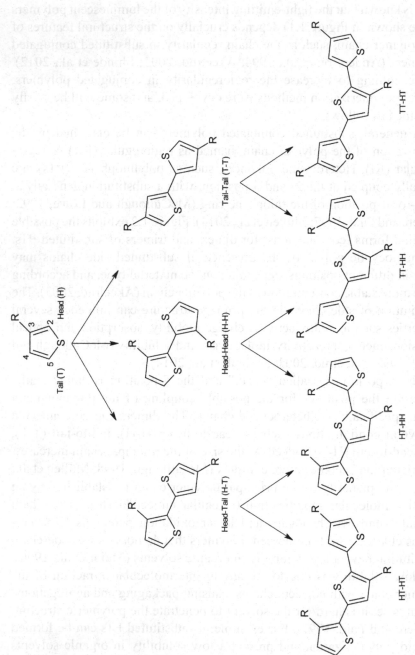

FIGURE 1.2 Regioisomerism orientation notations for 3-substituted polythiophenes. Reproduced with permission from Müllen et al. © 2014 The Royal Society of Chemistry.

improvement in the polymer solubility is observed when there is some substitution in the 3- or 4-position of the thiophene ring (Maior et al., 1990; McCullough, 1998; Müllen et al., 2014; Koeckelberghs et al., 2006).

1.2 ADDITION POLYMERIZATION

The addition polymerization method consists of chain-growth by a repetition of monomeric units. The mechanism involves the creation of a reactive intermediate, followed by the propagation of the monomers. That process originates from an addition polymer and does not result in by-products (Mark, 2011). Usually, the used monomers are those with unsaturation in their structure. Examples of some addition polymers (common polymers or some commodities polymers) are polyethylene (PE), polyvinyl chloride (PVC), polystyrene (PS), polyvinyl acetate (PVA), and polyacrylonitrile (PAN) (Davidson, 2008; Mark, 2011; Fleming et al., 2014).

1.2.1 OXIDATIVE POLYMERIZATION

Oxidative polymerization is one of the most used and known methods of addition. Sugimoto et al. (1985) were the pioneers of this type of oxidative cationic polymerization, practiced in the synthesis of some PTs, polypyrrole (PPy), and polyfuran (PF) (Yoshino et al., 1984). This route includes the addition of monomer dissolved in an aqueous medium or organic solvent. During this process, the addition of an oxidant agent is done, and a reaction between the thiophene ring and the oxidant agent occurs, producing a thiophene radical-cation (Figure 1.3). This radical-cation reacts with another one preferentially at 2,5-position of the ring. As this happens n times, a conjugated chain is formed (Barbarella et al., 1996; Fichou, 1999). The molar ratio of 1 mol of monomer to 4 mols of catalyst is set at constant stirring and room temperature (He et al., 2007; Müllen et al., 2014; Sadhanala et al., 2015). The commonly used catalyst is iron (III) chloride ($FeCl_3$) (Sugimoto et al., 1985). However, the addition polymerization can be performed by using other strong oxidant agents, such as molybdenum (V) chloride ($MoCl_5$), ruthenium (III), chloride ($RuCl_3$), or ammonium persulfate ($(NH_4)_2S_2O_8$) (Yoshino et al., 1984). In this regard, a dry reaction system should be used because the catalyst cannot undergo an oxidation reaction with the environment before having contact with the monomers. Torres and Balogh (2012) proposed how the oxidative cation reaction mechanism occurs.

The most accepted mechanism for oxidative polymerization suggests that a radical-cation oxidizes the thiophene derivative monomer at 2-position (most likely to occur oxidation) or at 5-position by reducing the catalyst ($Fe^{+3} \rightarrow Fe^{+2}$) (Andersson et al., 1994; Müllen et al., 2014). After, the coupling between the two radicals-cations occurs, providing the polymer chain growth (Torres and Balogh, 2012).

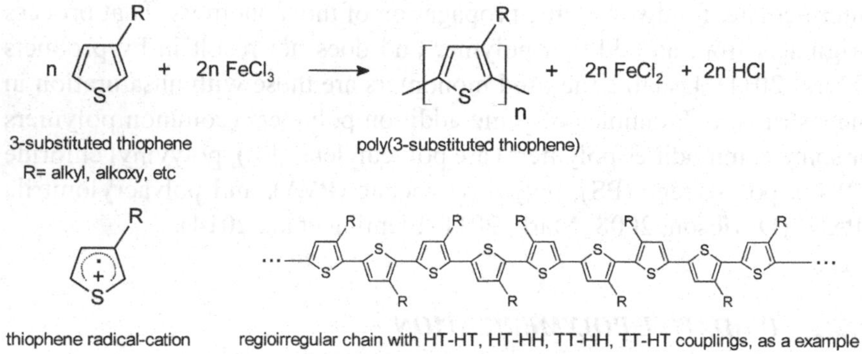

3-substituted thiophene
R= alkyl, alkoxy, etc

poly(3-substituted thiophene)

thiophene radical-cation   regioirregular chain with HT-HT, HT-HH, TT-HH, TT-HT couplings, as a example

FIGURE 1.3 Radical-cation coupling mechanism with high oxidation conditions for 3-substituted polythiophenes.

For the oxidative cationic mechanism, the possible couplings are H-H, T-T, and H-T. In the condition where the H-T coupling fraction is relatively low, the polymers are classified as RI polymers. These, with regularity lower than 70%, are known to have a low molar mass, high polydispersity, and particular optical properties due to the polymeric structure present in the material (Levesque and Leclerc, 1995; Stein et al., 1995; McCullough, 1998; Ohshita et al., 2009). Also, the influence of different types of side groups has been reported on the structural, optical, and thermal properties of some 3-substituted thiophene derivatives, such as the poly(3-hexyl-thiophene) (P3HT), poly(3-hexyloxythiophene) (P3HOT), and poly(3-phenylthiophene) (P3PhT), all synthesized by oxidative polymerization (He et al., 2007).

Several studies indicate that a slow addition of oxidant agents in the reaction produces more RR polymers with H-T coupling rates > 70% as compared to the instantaneous addition of the catalyst (Maior et al., 1990; Andersson et al., 1994; Hu and Xu, 2000). Moreover, for a low concentration of oxidant agents, the mechanism involves reactions between radical-cations and neutral molecules. Also, when the addition of a catalyst is slow, the coupling between oligomers and monomers is favored, rather than the linkage between oligomers themselves, thus increasing the probability of H-T structures (Andersson et al., 1994; McCullough, 1998; Torres and Balogh, 2012; Hu and Xu, 2000). In the case of instantaneous catalyst addition, most of the monomers are oxidized at the same time, leading to several radical-cations that immediately result in the coupling of T-T and H-T structures (Torres and Balogh, 2012).

A regioselective polymerization can therefore be reached by slow and controlled addition of a catalyst, such as $FeCl_3$. Andersson et al. (1994) reported a 94% H-T coupling in the 3-(4-octylphenyl)thiophene synthesis. Indeed, it has been shown that the chemical kinetics of addition polymerization are easier to control at a lower temperature (-20 to $0°C$), producing more organized polymers. It is noticed that the lesser the reaction temperature is, the more selective will be the coupling between the repeating units (Torres and Balogh, 2012).

Bizzarri et al. (1995) presented an oxidative polymerization for ester-functionalized poly(3-alkylthienylene)s using $FeCl_3$ as a catalyst and chloroform and nitromethane as solvents, i.e., a modified synthetic rote based on the Sugimoto method. The reactions in chloroform produced polymers with a high molecular weight, but an incomplete conversion of monomers. In contrast, the addition of nitromethane (CH_3NO_2) to the reaction system provided a complete polymeric conversion. However, the obtained polymers had a lower molecular weight when compared to those synthesized in chloroform. In this case, the total transformation of the monomer can be assumed as a better $FeCl_3$ dispersion instead of an increased number of initiator centers in reaction.

In general, after the reaction time, the PTs purification process occurs, which is done by precipitation in methanol, cooling for 24 hours, filtration, and polymer drying. Such a process can cause dedoping. An undoped polymer can be obtained by washing it in methanol or another non-solvent until a colorless solution is succeeded (Sadhanala et al., 2015). Similarly, a neutral state can be obtained by washing with hydrazine hydride (Ohshita et al., 2009; Hu and Xu, 2000). Soxhlet extraction is often used to remove possible impurities, such as oligomers (Andersson et al., 1994).

The advantages of addition polymerization over condensation polymerization are that the monomer does not require high purity, and there is no need for entirely inert systems and cryogenic temperatures. Furthermore, the reagents are of low cost compared to the other polymerization methods (Schlüter et al., 2012). However, the reproducibility of the addition polymerization is quite challenging, as discussed in the literature (Chen et al., 1996). Remarkable differences in molecular weight (ranging from 54 to 122 kDa) and remaining $FeCl_3$ amount as an impurity in the polymer were reported (Fichou, 1999).

1.3 CROSS-COUPLING POLYCONDENSATION

Condensation polymerization comprises step-growth or chain-growth of polymers by eliminating small molecules prevenient of monomeric precursors (Yokoyama and Yokozawa, 2009). This method has been used to prepare engineering polymers such as polylactic acid (PLA) (Ajioka et al., 1995), poly(ethylene terephthalate) (PET) (Dröscher and Wegner, 1978), polyamide (Ziabicki and Kedzierska, 1962), polycarbonate (PC) (Lee, 1964), and polyurethane (PU) (Gaboriaud and Vantelon, 1982). However, polycondensation catalyzed by transition metal complexes, such as nickel and palladium, is an important strategy to synthesize conjugated polymers. Three essential steps are involved in this type of polymerization: (i) intramolecular transference of the catalyst to the monomer in additive oxidation, (ii) transmetalation of the monomeric coupling partner to the catalyst/monomer complex, and (iii) the reductive elimination of the catalyst and the polymer formation after several cycles; and regeneration of the catalyst to the process.

Several transition metal-catalyzed methods have been employed to prepare PTs, poly(p-phenylene)s (PPP)s, poly(p-phenylene vinylene)s (PPV)s, and polyfluorenes (PFO)s. In general, PTs have been synthesized by nickel-catalyzed Grignard's metathesis (GRIM) method, whereas the palladium-catalyzed Suzuki-Miyaura (SM) coupling polymerization is used for PPPs, PFOs, and PPVs. Palladium-catalyzed Stille reaction is also a crucial synthetic route for PTs and copolymers based on PTs (Xu et al., 2014; Truong et al., 2015). A summary of these methods will be presented in the next subsections.

1.3.1 GRIGNARD'S METATHESIS (GRIM)

Kumada and co-workers were the first, in the 1980s, to develop polymer synthesis by the condensation method by using nickel as a catalyst (Tamao et al., 1982). The fact that nickel has a lower ionic volume than palladium results in more repeated units involved in the catalyst, which leads to polymers with higher degrees of regioregularity as compared to a synthesis that employed the palladium as a catalyst (Müllen et al., 2014). RR polymers, i.e., high H-T coupling rates, exhibit a higher molar mass, lower bandgap, superior conductivity, and higher luminescence as compared to RI polymers (Stein et al., 1995; McCullough, 1998; Sheina et al., 2005). RR polymers can be produced by condensation polymerization, where the polymeric chain growth is accomplished in steps generating by-products.

GRIM, via transition metal-assisted polymerization (Ni as a catalyst), has been recognized as a versatile method to produce 3-substituted RR polymers, such as PTs (Figure 1.4(a)) (Sheina et al., 2005). The Grignard reagent used in this type of polymerization has the general formula RMgX, where R is an alkyl or aryl group, and X is a halogen (Corriu and Masse, 1972). Usually, the halogens used are chlorine and bromine because they are more reactive than others, as the iodine, for example. They are particularly useful nucleophiles reacting with electrophiles (Corriu and Masse, 1972).

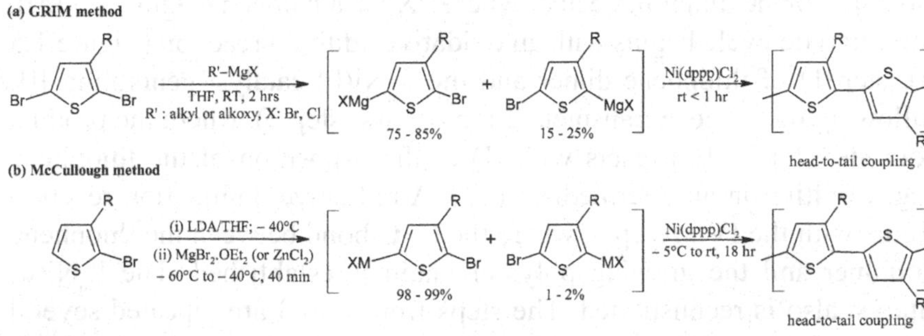

FIGURE 1.4 Common synthetic procedures for regioregular poly(3-alkylthiophenes). (a) Grignard Metathesis method. (b) McCullough method. Adapted from Müllen et al., (2014) – Reproduced by permission of The Royal Society of Chemistry.

The GRIM mechanism differs from the others, as the modified Kumada route by McCullough (see Figure 1.4(b)), because it does not require cryogenic temperatures (McCullough and Lowe, 1992; Sheina et al., 2005; Schlüter et al., 2012). The classical Kumada coupling is a type of cross-coupling reaction used for generating C-C bonds by the reaction of a Grignard reagent and an organic halide.

GRIM demand a dry solvent and is carried out in the presence of Ni as an organometallic catalyst. The used commonly catalysts are Ni(II) complexes, such as [1,3-bis(diphenylphosphino)propane]nickel(II) dichloride (Ni(dppp)Cl$_2$), [1,2-bis(diphenylphosphino)ethane]nickel(II) dichloride (Ni(dppe)Cl$_2$), [1,1'-bis(diphenylphosphino)ferrocene] nickel(II) dichloride (Ni(dppf)Cl$_2$), and bis(triphenylphosphine)nickel(II) dichloride (Ni(PPh$_3$)$_2$Cl$_2$), in the presence of the GRIM reagent known as alkyl/aryl magnesium halide (Iovu et al., 2005; Stefan et al., 2009). The substituted monomer is brominated (or iodized) at 2-position and 5-position of the thiophene ring. Afterward, transmetalation occurs between the GRIM reagent and the brominated monomers to form a mixture of isomers (see Figure 1.4(a)) (McCullough, 1998). Between isomers, the 2-bromo-5-bromomagnesium[3-alkylthiophene] is more stable and is formed with a greater molar percentage than the 5-bromo-2-bromomagnesium[3-alkylthiophene], for instance.

The monomer modified by the Grignard reagent (I) attacks the nickel catalyst twice. Therefore, the intramolecular transfer results in the formation of the initial organometallic catalyst (II), the L$_n$Ni(0) (Ln are ligands), and a thiophene dimer molecule, where "X" is a halogen (Figure 1.5 (a)). The catalytic cycle begins with an oxidative addition reaction (Figure 1.5 (b), step 1) of thiophene dimer and the L$_n$Ni(0) catalyst generating III. Following the cycle, a transmetalation occurs (step 2), where the product from step 1 (the III) reacts with (I) at the 5-position of the thiophene ring, resulting in an intermediate (IV). A reductive elimination reaction occurs with the IV in step 3, where the C–C bond between the thiophene monomer and the growing polymer chain is established. The L$_n$Ni(0) catalyst also is reconstituted. The steps from 1 to 3 are repeated several times to form a RR polymeric chain. Interestingly, Iovu et al. (2005) synthesized RR poly(3-alkylthiophene) by quasi-"living" chain-growth mechanism instead of a step-grown process. Kinetic studies revealed that the molecular weight is a function of the molar ratio of the monomer to the nickel initiator.

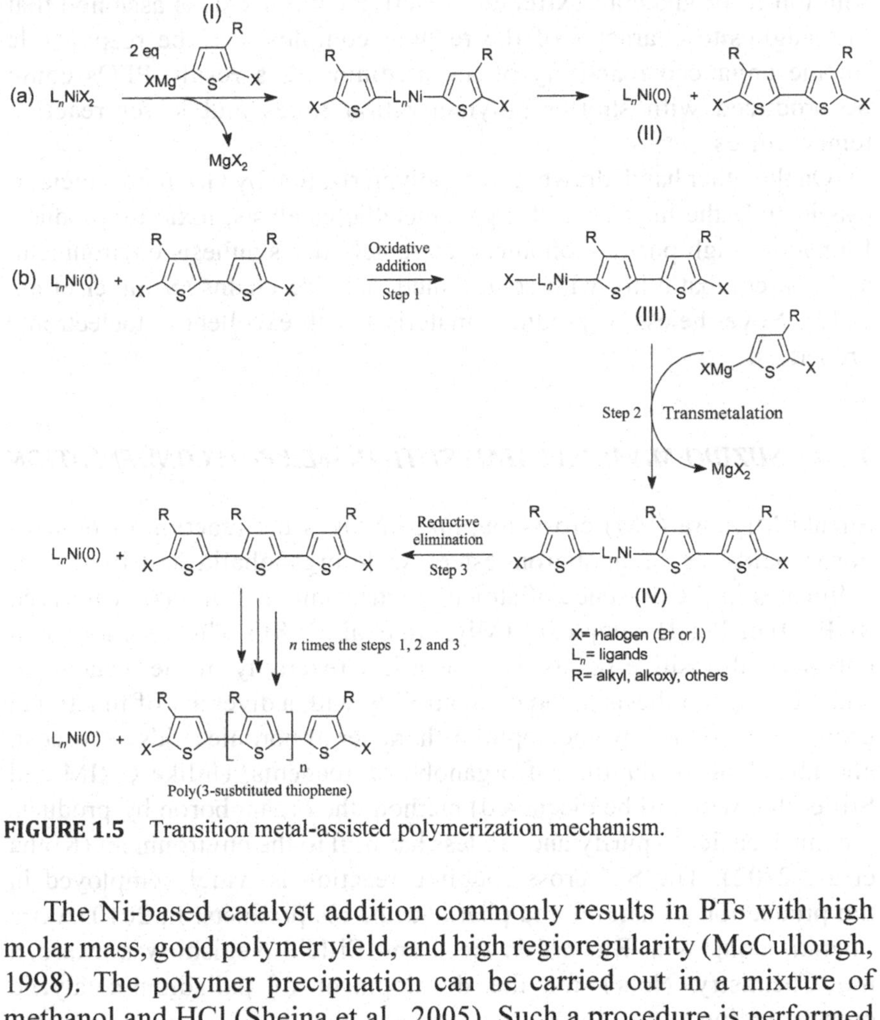

FIGURE 1.5 Transition metal-assisted polymerization mechanism.

The Ni-based catalyst addition commonly results in PTs with high molar mass, good polymer yield, and high regioregularity (McCullough, 1998). The polymer precipitation can be carried out in a mixture of methanol and HCl (Sheina et al., 2005). Such a procedure is performed to cease the reactivity of the GRIM reagent in the reaction. Furthermore, soxhlet extraction is additionally executed to remove possible impurities.

The Grignard's method applicability has been expanded through the "turbo-Grignard" reagents (i-PrMgCl^{-}2Li^{+}) by adding lithium chloride to a Grignard reagent. This approach has been used to synthesize a variety of aryl and heteroaryl conjugated polymers (Stefan et al., 2009). Turbo-Grignard breaks the polymeric aggregates formed in solution to produce an extremely reactive complex. It is assumed that the magnesite character of the reagent complex may be responsible for the enhanced reactivity of the medium. As a result, PFOs could be produced with shorter polymerization times and lower reaction temperatures.

On the other hand, drawbacks of polymerization by Grignard's metathesis include the high cost of organometallic catalysts, toxic by-products formation, high purity monomers, extremely dry synthesis environment, and non-compatibility with some functional side chains (Schlüter et al., 2012). Nevertheless, it produces materials with excellent optoelectronic properties.

1.3.2   SUZUKI-MIYAURA CATALYST-TRANSFER POLYCONDENSATION

Suzuki-Miyaura (SM) cross-coupling involves the reaction of organo-boron acids or organoboron esters with organohalides, triflates, or sulfonates in the presence of stoichiometric amounts of bases catalyzed by Pd(0) or Pd(II) complexes (Miyaura et al., 1981). This reaction presents several positive points, such as a low reactivity of the reagents to water during synthesis, good compatibility with a diversity of functional groups (due to the low nucleophilic character of boronic acids), low cost, and abundant availability of organoboron reagents. Unlike GRIM and Stille (that will still be discussed) method, the organoboron by-products are much easier to purify and are less harmful to the environment (Kotha et al., 2002). The SM cross-coupling reaction is widely employed in the production of natural and pharmaceuticals products. In 2010, Akira Suzuki, along with Richard F. Heck and Ei-Ichi Negishi was awarded the Chemistry Nobel for the development of palladium-catalyzed cross-coupling.

The catalyst cycle for the SM cross-coupling reaction is shown in Figure 1.6. The cycle begins with the oxidative addition of an aryl halide reagent to palladium species. Meanwhile, organoboron reagent reacts with the base to produce a borane species; afterward, it is transmetalated onto palladium catalyst/aryl halide species. The reductive elimination step generates the C–C (or R–R') bond and recycles the palladium catalyst (Varnado and Bielawski, 2012).

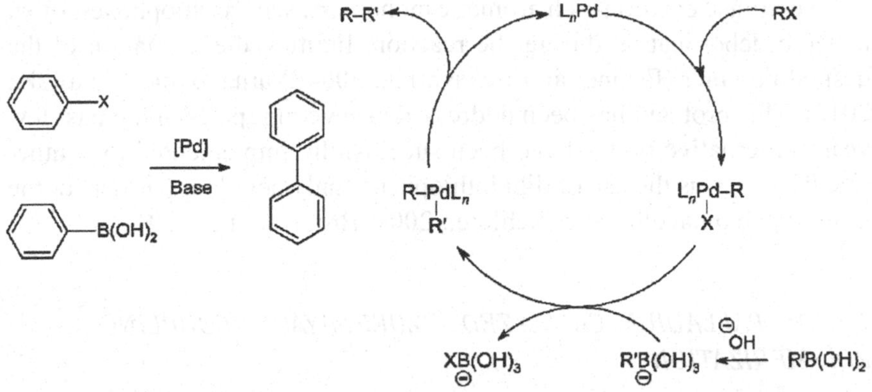

FIGURE 1.6 The Suzuki reaction (left) and the catalyst cycle of the Suzuki reaction (right). Reprinted from Varnado and Bielawski (2012), with permission from Elsevier.

The SM polycondensation implicates in the reaction of bifunctionalized monomers. Depending on the monomeric precursors, this route can be performed by combining AB-type monomers or AA-BB type monomers. AB-type polymers are synthesized by associating boronic acids or ester on one side of the monomer and halides (I and Br) on the other. AA-BB polymers are formed when bifunctionalized monomers bearing organoboron acids (or esters) react with dihalides monomers, as shown in Figure 1.7 (Schlüter, 2001). The SM method has been successfully used in the preparation of PPPs, PFOs, and PPVs, producing polymers with high molecular weight, low polydispersity, and high yields (Yokoyama et al., 2007; Yokoyama and Yokozawa, 2009; Varnado and Bielawski, 2012; Zhang et al., 2016).

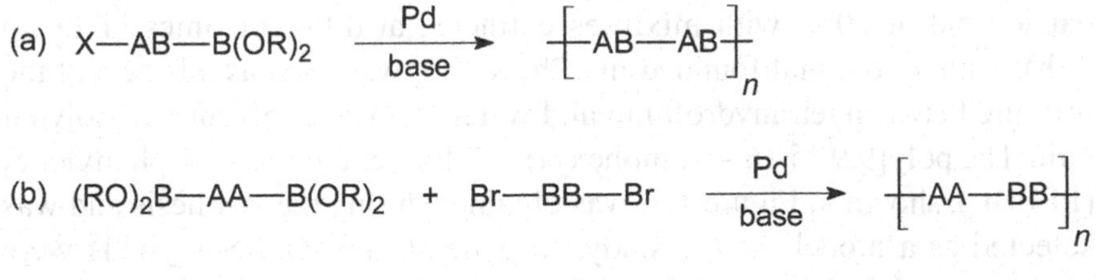

where, AB is monomer 1; AA is monomer 2; BB is monomer 3;

X is a halogen atom: Br, I or TfO ; R is an alkyl group or hydrogen atom

FIGURE 1.7 Representations for Suzuki polycondensation: (a) AB-type polycondensation; (b) AA-BB-type polycondensation.

In contrast, electron-rich aromatic monomers, such as thiophenes, often undergo deboronation during the reaction, limiting the formation of the desired polymer (Espinet and Echavarren, 2004; Varnado and Bielawski, 2012). This problem has been addressed in several reports in the past few years. Alternative routes have been successfully implemented to synthesize PTs, such as the use of dihalothiophene (only coupling partner) or the more stable pinacolborane (Schlüter, 2001; Brouwer et al., 2011).

1.3.2.1 *PALLADIUM-CATALYZED SUZUKI-MIYAURA COUPLING POLYMERIZATION*

The ligand and the oxidative state of palladium catalysts have an essential role in SM polycondensation. Both influences directly in the yield, polymeric conversion, kinetics, and polydispersity of the final polymer. The latter has been especially tricky to control and since narrow polydispersity is desirable for any polymeric synthesis.

Many palladium complexes have been explored over the years. Here, some examples are given to clarify the relationship between Pd/ligand in the success of SM coupling polymerization. Molina et al. (2009) investigated the processes involved in the synthesis of the poly(2,7-fluorene-*alt*-1,4-phenylene) (PFP) – a polymer that presents high blue luminescence. The polymerization degree and several chemical properties were studied using a model reaction varying the palladium catalysts. For this, the authors make use of four commercially available palladium catalysts with different oxidation states, configurations, and types of ligands. One Pd(0) species – (*1*) [Pd(PPh$_3$)$_4$], and three Pd(II) species – (*2*) (*E*)-[Pd(PPh$_3$)$_2$Cl$_2$] , (*3*) (*Z*)-[Pd(dppe) Cl$_2$] and (*4*) (*Z*)-[Pd(dppf)Cl$_2$]. The reactions were carried out at 80°C with mixtures extracted at different times (1 up to 72h), with an optimal founded in 12h. K$_2$CO$_3$ was used as a base and the mixture between tetrahydrofuran and water (1:1) as a solvent/non-solvent pair. The poly[9,9-bis(6'-bromohexyl)-2,7-fluorene-co-alt-1,4-phenylene] (PFPBr$_2$), shown in Figure 1.8, was obtained during the synthesis and was selected as a model for the study. In general, the synthesis yields were close in number, between 77 and 86%, and the degree of polymerization ranged from 23 to 44. When the catalyst (*4*) was used, PFPBr$_2$ with the highest degree of polymerization was obtained, although with the highest and broader polydispersity. The last one was attributed to an increment of the chain (monomer units) when compared to the other polymers produced by the catalysts (*1*), (*2*), and (*3*). In conclusion, a correct selection of a catalyst resulted in a polymer with a well-defined M_w and desired chain length.

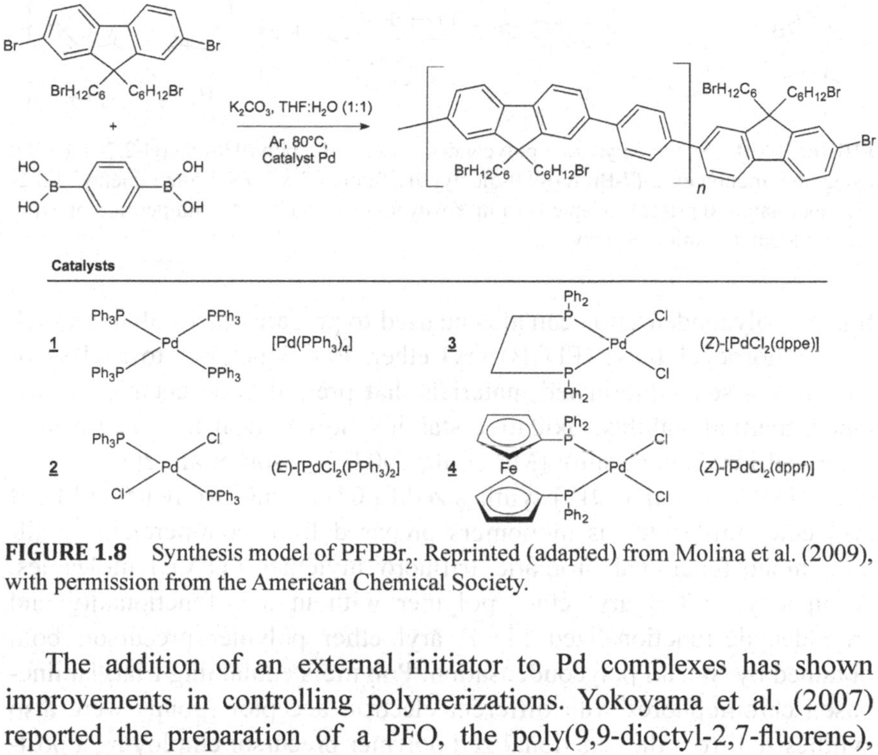

FIGURE 1.8 Synthesis model of PFPBr$_2$. Reprinted (adapted) from Molina et al. (2009), with permission from the American Chemical Society.

The addition of an external initiator to Pd complexes has shown improvements in controlling polymerizations. Yokoyama et al. (2007) reported the preparation of a PFO, the poly(9,9-dioctyl-2,7-fluorene), applying that idea (see Figure 1.9). The polyfluorene was synthesized in THF using Na$_2$CO$_3$ as a base, initiator bromo(tri-tert-butylphosphine) palladium(I) dimer (*t*-Bu$_3$PPd(Ph)Br), and a Pd(II) complex originated from the oxidative addition of bromine to Pd(0). The reactions were carried out at 80°C for 24h under argon atmosphere. PFOs presented low polydispersity (M_w/M_n) ranging from 1.33 to 1.39 with molecular weight (M_n) between 7.7 and 17.7 kDa. The phenyl group (Ph) of the initiator was found to be the end unit of polymer chains. The conversion of the monomer was linearly proportional to the size of the polymer chains, which suggests that the mechanism occurred in the form of chain-growth and a controlled manner. Zhang et al. (2016) reported the synthesis of AB-type fluorenes monomers by employing tri-tert-butylphosphine (t-Bu$_3$P) ligand associated with several external initiators. As a result, polymers with narrow polydispersity and well-controlled functional end groups were obtained.

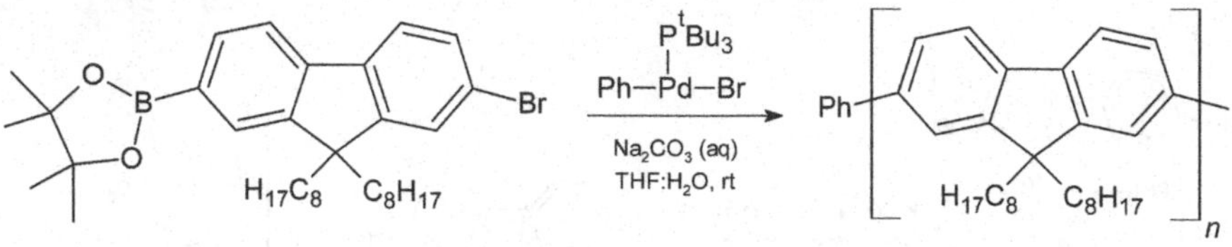

FIGURE 1.9 Suziki-Miyaura polycondensation of poly(9,9-dioctyl-2,7-fluorene) using the monomer 2-(7-Bromo-9,9-dioctyl-9H-fluoren-2-yl)-4,4,5,5-tetramethyl-1,3,2-dioxaborolane. Reprinted (adapted) from Yokoyama et al., (2007), with permission from the American Chemical Society.

Suzuki polycondensation can also be used to prepare other polymers such as perfluorocyclobutyl (PFCB) aryl ether. PFCB belongs to a class of amorphous semi-fluorinated materials that present good chemical resistance, thermal stability, oxidative stability, low optical loss, and a high electrical insulating ability (Ma et al., 2002; Neilson et al., 2007; Wu et al., 2015). Wu et al. (2015) synthesized PFCBs using two reactive PFCB aryl ether molecules as monomers prepared from commercially available mono-functional aromatic trifluorovinylether (TFVE) molecules. A biphenyl PFCB aryl ether polymer without any functionality and an aldehyde-functionalized PFCB aryl ether polymer precursor, both obtained by Suzuki polycondensation. Polymers containing triarylamine-based chromophores with different electron acceptor groups were also synthesized from the functionalized polymer precursor employing a post-polymerization modification procedure via Knoevenagel condensation (this method will be presented in 1.3.5 subsection). The polymers showed high solubility and good thermal stability, and when diluted in chloroform, exhibited significant contrasts in color (yellow, purple, blue, and orange). In summary, Suzuki polycondensation is also suitable for preparing and developing new functional PFCB aryl ether polymers.

It is known that polymers with a high molecular weight, via the conventional Suzuki route, can be obtained, although with high synthesis times. However, different strategies have been developed to decrease reaction time. Gao et al. (2014) used the ultrasound-assisted Suzuki coupling reaction to synthesize polydihexylfluorene. A reaction scheme is shown in Figure 1.10. A polymer with a high molecular weight (39.100 g/mol), soluble in a common organic solvent, was obtained in a short synthesis time (20 min) under optimized condition. An ultrasonic irradiation approach is a useful tool for Suzuki coupling polymerization on a large scale and highly desirable for commercial applications.

Nevertheless, it required strict control of ultrasound irradiation power because a polymer chain degradation can occur under strong ultrasound input. Gao et al. (2014) concluded that the ultrasound-assisted synthesis could significantly accelerate the Suzuki coupling reaction and obtain conjugated polymers with high quality with remarkably short reaction time. The success in increasing the reaction rate was attributed to surface morphology changes and an enlarged surface area of the palladium catalyst.

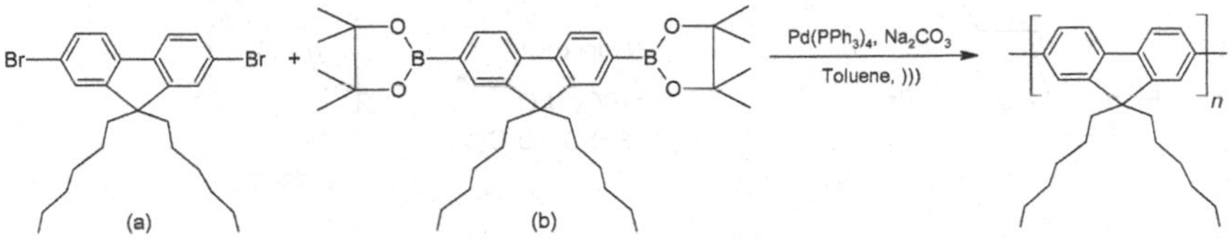

FIGURE 1.10 Palladium-catalyzed Suzuki coupling reaction under ultrasound irradiation for synthesis of polydihexylfluorene using (a) 2,7-dibromo-9,9-dihexyl-9H-fluorene, and (b) 2,2′-(9,9-dihexyl-9H-fluorene-2,7-diyl)bis(4,4,5,5-tetramethyl-1,3,2- dioxaborolane) as reagents. Reprinted (adapted) from Gao et al., (2014), with permission from Elsevier.

The nickel-catalyzed Suzuki cross-coupling has also been explored. Qiu et al. (2016) demonstrated a controlled Ni-catalyzed polyconden-sation. The conversion and selectivity of a pinacolborane thiophene monomer called thiophene-2-boronic acid pinacol ester (ThBPin) were studied with different loadings of nickel catalysts such as Ni(PPh$_3$)IPrCl$_2$, Ni(dppp)Cl$_2$ and Ni(1-Naph)(PCy$_3$)$_2$Br in comparison to a precatalyst: the [1,3-bis(2,6-diisopropylphenyl)imidazole-2-ylidene](3-chloropyridyl) palladium(II) dichloride (PEPPSI-IPr). Loadings of 5% mol led to high ThBpin conversion and selectivity greater than 90% for all Ni catalysts. However, a noticeable decrease in conversion and selectivity was observed when 1 mol % Ni(1-Naph)(PCy$_3$)$_2$Br. On the other hand, high conversions but lower selectivity were noticed to PEPPSI-IPr precatalyst at 1% and 5% mol loading. The homopolymers and the diblock thiophene copolymer (using monomers 1, 2, and 3, and the Ni catalysts – Figure 1.11) presented yields ranging from 52 to 79% with polydispersity between 1.13 and 1.63

and a molecular weight from 4.5 to 74.4 kDa. The polymerization of the monomer (indicated as (1) in Figure 1.11) using PEPPSI-IPr presented a 20% yield, a molecular weight of 5.5 kDa, and a dispersity index of 1.28. The authors have then demonstrated the versatility of the SM cross-coupling method by controlling the synthesis of ester-functionalized conjugated polymers.

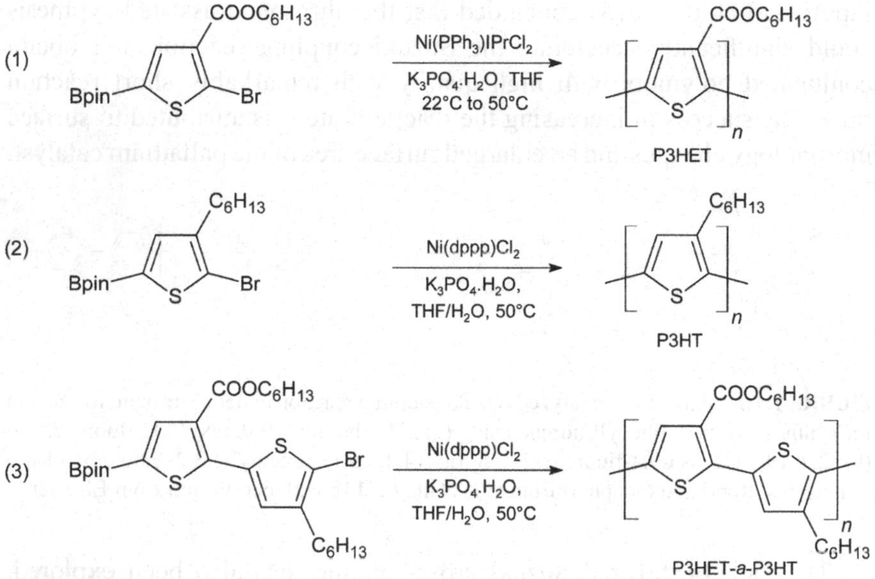

FIGURE 1.11 Poly(3-hexylesterthiophene) (P3HET) and poly(3-hexylthiophene) (P3HT) homopolymers were prepared using a Ni-catalyzed Suzuki coupling reaction. An alternating copolymer consisting of P3HET and P3HT was also synthesized. Reprinted from Qiu et al., (2016), with permission from the American Chemical Society.

Another approach employed to reduce the Suzuki reaction time is microwave-assisted polymerization, where reactions can be performed in minutes. Indeed, microwave irradiation is effective in shortening reaction times, which is suitable for increasing reaction yields and obtaining high molecular weight materials, while the amount of by-products is reduced (Melucci et al., 2002; Nehls et al., 2004; Zhang et al., 2013; Metzler et al., 2015; Vázquez-Guilló et al., 2018).

The technique principle is based on the heat transfer achieved by dielectric heating, which, in turn, is primarily dependent on the ability of the solvent or reagent to absorb microwave energy. Therefore, there is a dissipation of induced energy in the form of dielectric loss and heat by molecular friction – the tendency of dipoles to follow the electric field (Larhed et al., 1996; Kappe and Dallinger, 2009; Mehtaa and Eycken, 2011; Baig and Varma, 2012; Sharma et al., 2018). However, microwave-assisted organic synthesis may present an inconvenience in some cases; insoluble gels are formed during the reaction and affect the yield of conjugated polymers (Kappe and Dallinger, 2009; Zhang et al., 2013).

Zhang et al. (2013) used the microwave-assisted Suzuki coupling reaction to prepare poly(9,9-dihexylfluorene)s (PDHFs). The authors systematically investigated the synthesis parameters, including the mode and microwave power, temperature and reaction time, species and catalyst concentrations, and solvents. PDHFs has been obtained with a high molecular weight (40,000 g/mol), good yield (65%), and high polydispersity (1.96) when a power of 150 W was applied in 14 min. The polymerization reaction carried out at 130°C using THF as a solvent and $PdCl_2(dppf)$ as a catalyst. Microwave power and reaction times greater than 150 W and 14 min, respectively, produced an insoluble gel-like product and, thus, decreasing the soluble polymer yields. Therefore, they are critical factors for synthesis success.

Despite these critical details, where both power and time control plays an essential role in the success of the synthesis, the great advantage of microwave irradiation is in reactional time, when compared to the conventional synthesis of conjugated polymers by palladium-catalyzed Suzuki polymerization reaction. The latter requires long reaction periods (dozens of hours) to obtain the desired product.

1.3.3 STILLE CROSS-COUPLING POLYCONDENSATION

Stille reaction involves cross-coupling of organotin reagents R"–Sn(R)$_3$, also known as organostannanes reagents (vinyl or aryl tin) with organohalides or sulfonate reagents R'–X catalyzed by palladium complexes to form a coupled product R'–R" and an undesirable organotin by-product XSn(R)$_3$ (Milstein and Stille, 1978, 1979; Stille, 1986; Farina et al., 1998).

The generic scheme of the Stille reaction using the Pd catalyst is shown in Figure 1.12(a). Notably, the organotin compounds are not sensitive to moisture or oxygen and present high compatibility with various functional groups, unlike Grignard reagents. Its general mechanism is depicted in Figure 1.12(b).

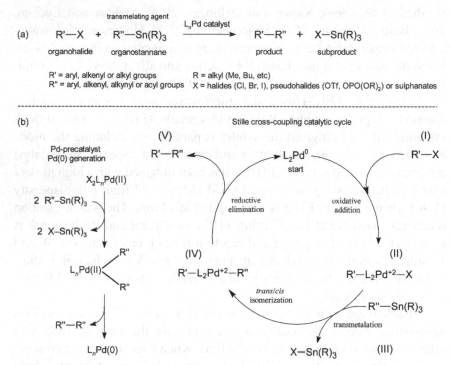

FIGURE 1.12   (a) A generic scheme of the Stille reaction. (b) The general mechanism for Stille catalytic cycle.

The catalytic cycle starts with the oxidative addition of (I) to a complex of palladium with a ligand (L_2Pd) to form a Pd(II) complex, such as the $R'-L_2Pd^{+2}-X$ intermediate (II). The organotin functionalized molecule (III) is then transmetalated to palladium (where the R" group of the organotin reagent replaces the halide anion on the palladium complex) forming R $-L_2Pd^{+2}-R"$ (IV). The last step (V) consists of the reductive elimination of L_2Pd of (IV), formation of the final coupled product R'–R", and the regeneration of the palladium catalyst to the cycle. The known order of the rate of transmetalation (ligand transfer) from a tin is alkynyl > alkenyl > aryl > allyl > alkyl. Several excellent reviews can be found in the literature (Stille, 1986; Mitchell, 1986, 1992; Farina 1996; Farina et al.,

1998; Genet and Savignac, 1999; Kosugi and Fugami, 2002; Espinet and Echavarren, 2004; Carsten et al., 2011; Cordovilla et al., 2015).

A general route to prepare aromatic polymers using palladium-catalyzed cross-coupling reactions with difunctional tin reagents was first introduced by Bochmann and Kelly (Bochmann and Kelly, 1989). The polymerizations of *para*-phenylenes were carried out in the temperature range between 130 and 165°C, employing N,N-dimethylacetamide as a solvent, and $Pd(PPh_3)_2Cl_2$ as a catalyst. The results suggested that structures based on rigid building units (internal $C{\equiv}C$ linkages) possess the least solubility and, correspondingly, the highest halide content and lowest molecular weight. On the other hand, the introduction of more flexible linkages ($-O-$ or $-CH_2-$) and the use of highly soluble precursors increases the solubility of the growing polymer chain during the coupling process, forming medium chains length, with a higher molecular weight.

Further advances followed when alternated copolymers of thiophene with a *para*-phenylene derivative were obtained by Stille coupling reaction using a common solvent (THF) and $Pd(PPh_3)_2Cl_2$ (Bao et al., 1993). The copolymer presented a high molecular weight (14,000 g/mol), solubility in different organic solvents (THF, chloroform, and dichloroethane), and good fusibility, allowing easy processing. The incorporation of flexible side chains in the benzene ring was fundamental in the methodology, improving the solubility and increasing the molecular size of the polymer chains. In summary, the tolerance to different substituents makes the Stille reaction extremely useful for synthesizing soluble and fusible conjugated polymers.

Stille polycondensation is also a pathway to synthesize RR PTs by incorporation of organotin and halides in the 2,5 positions of the thiophene ring. McCullough et al. (1997) reported the preparation of RR oxazoline PT with approximately 100% HT-HT coupling and an 84% yield by using a CuO-modified Stille route. Lère-Porte et al. (1999) investigated the synthesis of a RR poly(alkylbithiophene) by an organometallic cross-coupling reaction using PPh_3 added to $Pd_2(dba)_3CHCl_3$ catalytic complex. The polymerization resulted in more than 90% HT-HT coupling, with a high molecular weight (34 kDa) and a low polydispersity (1.2).

A modification of Stille polycondensation consists of the microwave irradiation employment that allows reactions to occur at higher temperatures than in conventional reflux systems and in less time. Tierney et al. (2005) reported the synthesis of soluble PTs via microwave-assisted Stille polycondensation. Several factors were investigated by the authors to improve the molecular weights of the polymers and reduce the reaction time required to obtain high molecular weights. Under optimized microwave-assisted reaction conditions and suitable catalyst systems, the resulting PTs presented higher molecular weights ($M_n > 15$ kDa) and lower polydispersity (≈ 2.0) than that PTs synthesized by reflux at 135°C in 24 h of reaction ($M_n = 13$ kDa and $M_w/M_n = 2.3$).

1.3.4   WITTIG'S AND HORNER'S REACTIONS

The Wittig reaction is known as one of the most versatile methods for alkene synthesis. It consists of a chemical reaction of an aldehyde or a ketone with phosphonium ylide compounds (often called a Wittig reagent) to form an alkene and phosphine oxide (Wittig and Geissler, 1953). The reaction was developed by Georg Wittig, who was a Nobel Prize laureate in 1979.

The general reaction (homogenous) is illustrated in Figure 1.13(a), where the C=O carbonyl group is converted to a C=C double bond (sp2 hybridization). The Wittig reaction mechanism involves two steps (Figure 1.13(b)). A reversible step which comprises the reaction of a nucleophilic phosphonium ylide with the electrophilic carbonyl carbon of an aldehyde (or a ketone), giving a dipolar intermediate specie: a betaine. An intramolecular reaction occurs between the oxygen and phosphorus of the betaine (via four-centered transition states), producing a four-membered oxaphosphetane ring (Vedejs and Peterson, 1994). As a second step, irreversible decompositions of oxaphosphetanes (*cis* and *trans*) occur. The oxaphosphetane is unstable and breaks down by a *syn*-cycloreversion process to form *Z*- and *E*-stereoselective alkenes and triphenylphosphine oxide as a by-product. (Restrepo-Cossio et al., 1997). Reactions of stabilized (non-stabilized) phosphonium ylides with aldehydes favors *E*-alkene (*Z*-alkene) products. However, the degree of stereoselectivity varies with the reaction conditions, the nature of the ylide and carbonyl components as well the substituents, as expected.

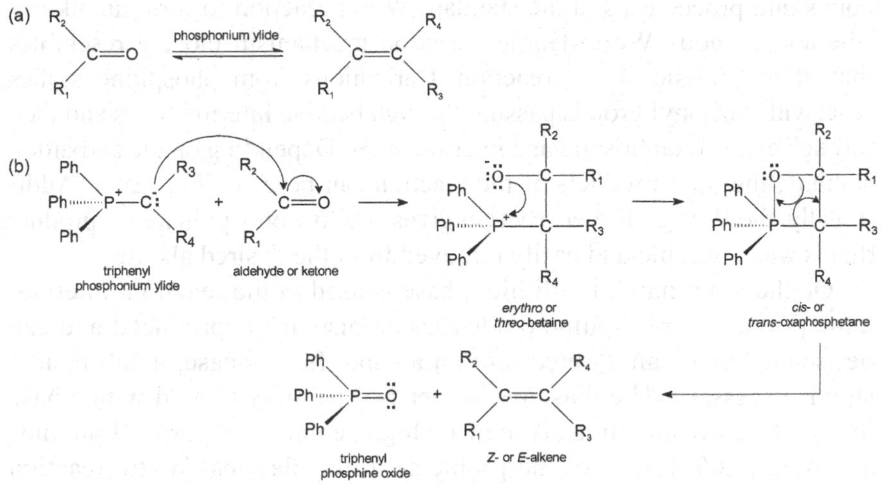

FIGURE 1.13 Wittig: (a) reaction scheme and (b) reaction mechanism. Reprinted (adapted) from Maryanoff and Reitz (1989), with permission from the American Chemical Society.

A variety of phosphorus reagents have been developed and applied for this transformation, yielding alkenes with different degrees of substitution. Depending on the nature of the reagent, the reaction is divided into three main groups (i) "classic" Wittig reaction of phosphonium ylides, (ii) Wittig-Horner reaction of phosphine oxide anions, and (iii) Horner-Wadsworth-Emmons reaction of phosphonate anions. Each of these reactions has its distinct advantages and limitations, and these must be considered when selecting the appropriate method for the desired synthesis (Edmonds and Abell, 2004).

A drawback of the Wittig reaction is that it is subject to steric hindrance. In general, the alkenes production from aldehydes (that have the least hindered carbonyl group) is higher than ketones, in which the carbonyl group is more hindered (Brown et al., 2008). A detailed account of the Wittig mechanism can be found in excellent reviews (Maryanoff and Reitz, 1989; Vedejs and Peterson, 1994, 1996; Pascariu et al., 2003).

Horner and co-workers (1958, 1959) were the first to change the standard Wittig reaction and describe the use of phosphine oxides in the preparation of alkenes. Such a modification allows for the removal of phosphorous as a water-soluble side product. If a base (e.g., KOH or NaH) is used to produce a phosphine oxide anion, the reaction with a carbonyl compound proceeds as in the standard Wittig reaction to form an alkene. The homogenous Wittig-Horner reaction mechanism closely resembles that of the classic Wittig reaction. Carbanions from phosphine oxides react with carbonyl groups passing through betaine intermediates and then through cyclic oxaphosphetane intermediates. Depending on the carbanion stability, the final products of the reaction can be Z- or E-alkenes. Additionally, the Wittig-Horner reaction gives rise to a phosphinate by-product that is water-soluble and easily removed from the desired alkene.

On the other hand, if a lithium base is used in the reaction, intermediate β-hydroxy phosphine oxide diastereoisomer is produced and can be isolated and transformed into an alkene. In this case, a subsequent step is necessary. The diastereoisomer is separately treated with a base to give the corresponding alkene with high geometrical purity (Edmonds and Abell, 2004). One of the problems in the classical Wittig reaction is the separation and purification of the alkene and phosphine oxide products. Phosphinate derivatives obtained from the Wittig-Horner reaction are all soluble in water; hence separation of the alkene is facile to achieve.

In 1961, Wadsworth and Emmons described the increased reactivity of phosphonate-stabilized carbanions with electron-withdrawing substituents (Wadsworth and Emmons, 1961). Treatment of phosphonate esters with strong bases produces carbanions for the reaction, which, followed by aldehydes or ketones, gives alkenes. Phosphonate-stabilized carbanions are more nucleophilic and, therefore, significantly more reactive than the phosphonium ylides employed in classical Wittig reaction. An advantage of using phosphonate-stabilized ylides instead of phosphonium salts (similar to what occurs in the Wittig-Horner reaction) is that the by-products, dialkyl phosphate salts, are water-soluble and therefore readily removed by aqueous extraction. Also, in contrast to phosphonium ylides, phosphonate-stabilized carbanions can be alkylated.

The mechanism for the Horner-Wadsworth-Emmons (HWE) reaction is as depicted in Figure 1.14. The reaction of phosphonate-stabilized carbanions with aldehydes (or ketones) form the oxyanion intermediates (IA and IB) under reversible conditions. Rapid decompositions of IA and IB into four-centered intermediates (IIA and IIB) provide Z- and E-alkenes.

The HWE reaction stereoselectivity depends on the nature of the phosphate utilized. It also results from both kinetic and thermodynamic

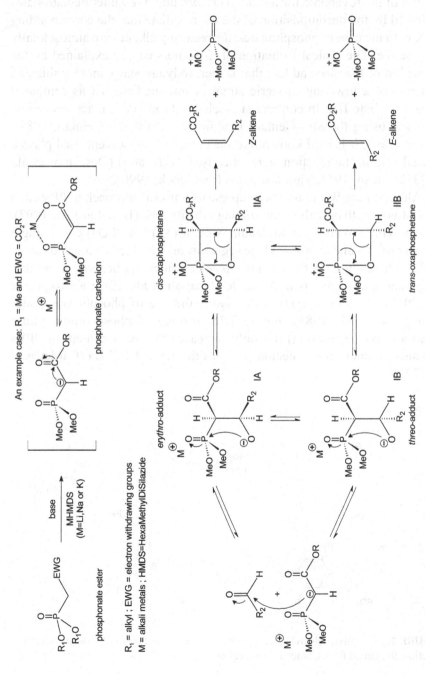

FIGURE 1.14 The Horner-Wadsworth-Emmons reaction mechanism.

control of the reversible formation of *erythro* and *threo* intermediate states followed by the decomposition of the intermediate into the corresponding alkene. In the case of phosphonates, the resulting alkenes are almost totally *E*-selective. The typical formation of *E*-isomers can be explained by the formation of the *threo* adduct that is thermodynamically more stable and in terms of a lowering of steric strain in intermediate IIB as compared to intermediate IIA. In contrast, a *Z*-selectivity in HWE reactions can be achieved using the Still-Gennari modification (Still and Gennari, 1983). Nevertheless, it is well known that the nature of the solvent used plays a crucial role in the reaction stereochemistry (*Z/E* ratio) (Descamps et al., 1973; D'Incan, 1977; Thompson and Heathcock, 1990).

Although the Wittig reaction represents a useful approach to PPVs and derivatives (with or without alternating substituents) (Brandon et al., 1997), only polymers with a low molecular weight (Mn < 10 kDa) containing a mixture of *cis*- and *trans*-vinylene segments are obtained (Yang and Geise, 1992). Depending on the organo optoelectronics application, a mixture of *cis* and *trans* units is undesirable (Jaballah et al., 2011; Karpagam et al., 2011). For circumventing this issue, the use of phosphorous ylides (Lampman et al., 1985), Figure 1.15, instead of phosphonium ylides (standard Wittig reaction) not only increases the *trans*-content in PPVs but also enhances the molecular weights far beyond 10 kDa (Cárdenas et al.; 2016).

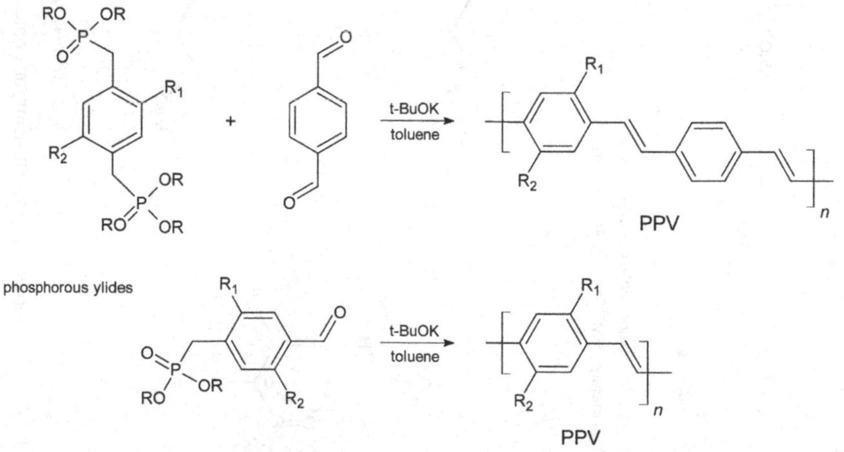

FIGURE 1.15   Poly(*p*-phenylene vinylene)s syntheses by Horner-Wadsworth-Emmons reaction. Reprinted from Cárdenas et al., (2016), with permission from authors.

Horner-based polycondensations have then been widely applied in the PPVs and derivates synthesis for the preparation of homopolymers and alternating copolymers (Lux et al., 1997; Pfeiffer and Hörhold, 1999; Suzuki et al., 2007; Wang et al., 2008; Laughlin and Smith, 2010). As aforementioned, the HWE approach involving the use of phosphonates is generally preferred over the phosphonium salts due to its higher reactivity of the ylide intermediate. The resulting polymers present higher degrees of polymerization and a predominately *trans* configuration of double bonds (Pfeiffer and Hörhold, 1999).

HWE reaction was used in the preparation of RR poly[(2-methoxy-5-alkyloxy)-1,4-phenylenevinylene]s (PPVs) from asymmetrically functionalized monomers of AB-type. X-ray diffraction data revealed that these RR PPVs have higher crystallinity when compared to RI PPVs in the solid-state (Suzuki et al., 2007). Also, highly RR poly[(2-methoxy-5-(3',7'-dimethyloctyloxy))-1,4-phenylenevinylene]s (MDMO-PPVs) were obtained by the HWE method and applied in photovoltaic devices (Tajima et al., 2008). The superior properties of RR MDMO-PPVs (compared to regiorandom MDMO-PPVs) were then validated by the improvement of the photovoltaic performance.

1.3.5   KNOEVENAGEL CONDENSATION POLYMERIZATION

Knoevenagel reaction is a classic organic synthesis for C=C bond formation, described by Emil Knoevenagel in the 1890s (Knoevenagel, 1898). Knoevenagel encloses the reaction (Figure 1.16) between a carbonyl compound (aldehyde or ketone) and an activated methylene group (catalyzed by an amine base, for instance), resulting in adduct formation with a carbon-carbon double bond (alkene) followed by water elimination (Tietze and Beifuss, 1992; Zerong, 2010). For the condensation between carbonyl groups and compounds with active methylene groups to form polymers, the reaction is referred to as Knoevenagel condensation polymerization.

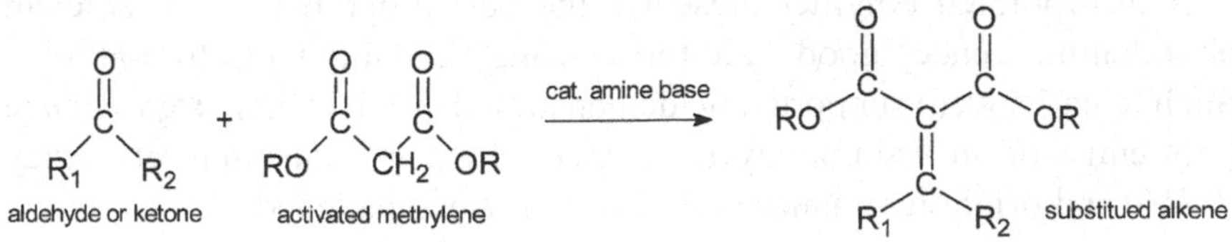

FIGURE 1.16   Knoevenagel condensation reaction.

Using at least one electron-withdrawing group, such as nitro ($-NO_2$), cyano ($-CN$), carbonyl ($C=O$), or sulfonyl (R_2SO_2) group, the methylene group can be activated. As a consequence, the compounds with any combination of these electron-withdrawing groups, adjacent to a methylene group, can act as nucleophiles, including malonic amides, malonic acid, malonic esters, cyanoacetamides, α-keto sulfones, among others (Zerong, 2010). This reaction can be catalyzed by weak amine bases, such as piperidine, pyridine, diethylamine, for instance.

The first step from the mechanism, shown in Figure 1.17(a), is the deprotonation of the activated methylene by a catalyst base (e.g., piperidine) to provide a resonance-stabilized enolate. The catalyst also reacts with the aldehyde or ketone to produce an iminium ion intermediate, which is then attacked by the enolate (Figure 1.17(b)). The base abstracts a proton from the intermediate compound generating another enolate, while the amine of the intermediate is protonated. Afterward, a rearrangement occurs that releases the amine base, regenerates the catalyst, and delivers the final α-β unsaturated carbonyl products (Knoevenagel, 1898; Gjiri, 2020). The reaction yield depends not only on the relative reactivity of the reactants, the strength of the bases employed, and the nature of solvents, but also on the removal of the formed water during the reaction (Zerong, 2010).

Knoevenagel condensation has been applied to the synthesis of soluble luminescent polymers. Several high electron affinity polymers for LEDs, cyano derivatives based on PPV, have been synthesized by Knoevenagel condensation (Moratti et al., 1995). The HOMO-LUMO gap could be tuned (between 3.5 and 1.8 eV) by the selection of the monomer units or the solubilizing groups.

Li et al. (1999) reported a conjugated polymer containing electron-withdrawing cyano and electron-rich groups (triarylamines). The synthesis was also carried out via Knoevenagel reaction between 1,4-bis(cyanomethyl)-2-[(2-ethylhexyl)oxy]-5-methoxybenzene and dialdehydes of two triarylamines: triphenylamine (TPA) and tetraphenyldiaminobiphenyl (TPD) as a hole injecting/transporting chromophores. In a comparative study, the TPA incorporated polymer presented the best properties; more efficient photoluminescence, good hole-transporting, and high electron-affinity. Such remarks were supported by demonstrated efficient and bright orange light emission in a single layer LED. Good external quantum efficiency (0.1%) and bright luminance of 2100 cd/m^2 were achieved.

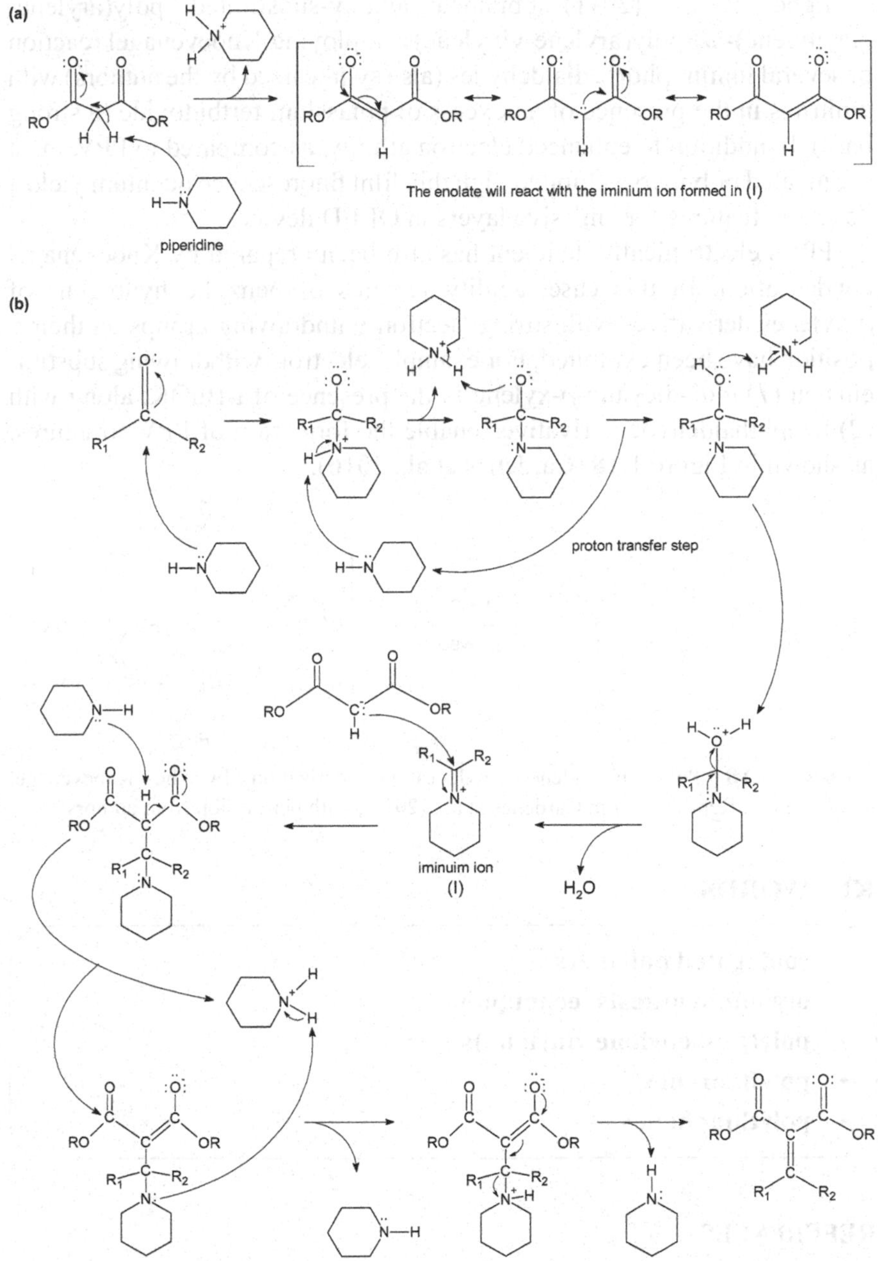

FIGURE 1.17 Knoevenagel condensation detailed mechanism. Reprinted from Gjiri (2020), with permission from the author.

Egbe et al. (2011) obtained alkoxy-substituted poly(arylene-ethynylene)-*alt*-poly(arylene-vinylene)s employing Knoevenagel reaction of several luminophoric dialdehydes (also synthesized by the authors) with dinitriles in the presence of an excess of potassium tertbutoxide (a strong base). In addition to enhanced electron affinity, as compared to PPV, most conjugated polymers exhibited high thin film fluorescence quantum yields, desirable features for emissive layers in OLED devices.

PPVs electronically deficient has also been prepared by Knoevenagel condensation. In this case, acidity features of benzylic hydrogens of *p*-xylenes derivatives with strong electron-withdrawing groups on their α position have been exploited. For example, electron-withdrawing substituents on (*1*) α,α'-dicyano-*p*-xylene in the presence of t-BuOK, along with (*2*) terephthaldehyde derivatives enable the formation of PPV structures, as shown in Figure 1.18 (Cárdenas et al., 2016).

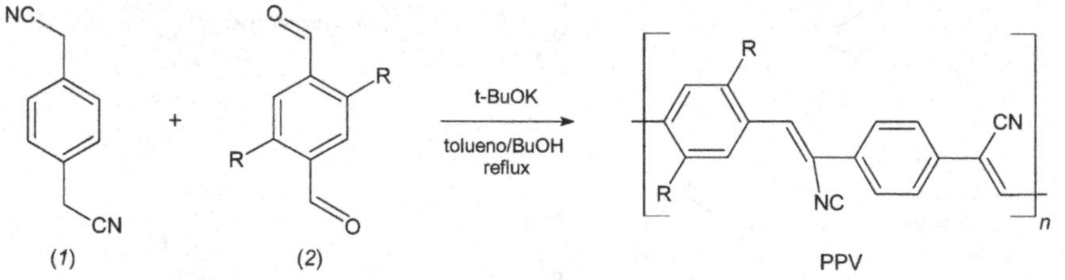

FIGURE 1.18 Poly(p-phenylene vinylene) polymerization by the Knoevenagel condensation. Reprinted from Cárdenas et al., (2016), with permission from authors.

KEYWORDS

- **conjugated polymers**
- **organic synthesis techniques**
- **poly(*p*-phenylene vinylene)s**
- **polyfluorenes**
- **polythiophenes**

REFERENCES

Ajioka, M., et al. (1995). The basic properties of polylactic acid produced by the direct condensation polymerization of lactic acid. *Bull. Chem. Soc. Jpn.*, *68*(8), 2125–2131.

Akcelrud, L., (2003). Electroluminescent polymers. *Prog. Polym. Sci.*, *28*, 875–962.

Andersson, M. R., et al., (1994). Regioselective polymerization of 3-(4-octylphenyl) thiophene with FeCl3. *Macromolecules*, *27*(22), 6503–6506.

Armour, M., et al., (1967). Colored electrically conducting polymers from furan, pyrrole, and thiophene. *J. Polym. Sci. Part A-1*, *5*, 1527–1538.

Baig, R. B. N., & Varma, R. S., (2012). Alternative energy input: Mechanochemical, microwave, and ultrasound-assisted organic synthesis. *Chem. Soc. Rev.*, *41*(4), 1559–1584.

Bao, Z., et al., (1999). Polymer light-emitting diodes: New materials and devices. *Opt. Mater.*, *12*, 177–182.

Bao, Z., et al., (1993). Synthesis of conjugated polymer by the Stille coupling reaction. *Chem. Mater.*, *5*(1), 2, 3.

Barbarella, G., et al. (1996). Regioselective oligomerization of 3-(alkylsulfanyl)thiophenes with ferric chloride. *J. Org. Chem.*, *61*(23), 8285–8292.

Bizzarri, P. C., et al., (1995). Ester-functionalized poly(3-alkylthienylene)s: Substituent effects on the polymerization with FeCl3. *Synth. Met.*, *75*(2), 141–147.

Bochmann, M., & Kelly, K., (1989). Palladium-catalyzed cross-coupling reactions with difunctional tin reagents: A general route to aromatic polymers. *J. Chem. Soc., Chem. Commun.*, 532–534.

Brandon, K. L., et al. (1997). Electroluminescent properties of a family of dialkoxy PPV derivatives. *Synth. Met*, *91*(1–3), 305–306.

Brouwer, F., et al., (2011). Using bis(pinacolato)diboron to improve the quality of regioregular conjugated copolymers. *J. Mater. Chem.*, *21*(5), 1582–1592.

Brown, W. H., et al., (2008). Aldehydes and Ketones. In: *Organic Chemistry* (5th edn.). Brooks/Cole Cengage Learning.

Cárdenas, J. C., Ochoa-Puentes, C., & Sierra, C. A., (2016). Phenylenevinylene systems: The oligomer approach. In: Yilmaz, F., (ed.), *Conducting Polymers* (pp. 223–240). InTech., Under CC BY 3.0 license. Available From http://dx.doi.org/10.5772/63394 (accessed on 17 June 2020).

Carsten, B., et al. (2011). Stille polycondensation for synthesis of functional materials. *Chem. Rev.*, *111*(3), 1493–1528.

Chen, F., et al., (1996). Improved electroluminescence performance of poly (3-alkylthiophenes) having a high head-to-tail (HT) ratio. *J. Mater. Chem.*, *6*(11), 1763–1766.

Cordovilla, C., et al. (2015). The Stille reaction, 38 years later. *ACS Catal.*, *5*, 3040–3053.

Corriu, R. J. P., & Masse, J. P., (1972). Activation of Grignard reagents by transition-metal complexes: A new and simple synthesis of trans-stilbenes and polyphenyls. *J. Chem. Soc., Chem. Commun.*, *3*, 144.

Davidson, C., (2008). Production of materials. In: *Chemistry Contexts* (p. 2). Pearson.

Descamps, B., et al. (1973). Mecanisme de la reaction de Horner-Emmons – II: Effet de solvant sur la stereoselectivite de la reaction d'aldehydes aromatiques et de phosphononitriles, *Tetrahedron*, *29*, 2437–2445.

D'Incan, E. (1977). Catalyse par transfert de phase et extraction par paires d'ions. stereoselectivite de la reaction de Horner-Emmons. *Tetrahedron*, *33*, 951–954.

Dröscher, M., & Wegner, G., (1978). Poly(ethylene terephthalate): A solid-state condensation process. *Polymer*, *19*(1), 43–47.

Edmonds, M., & Abell, A., (2004). The Wittig reaction. In: *Modern Carbonyl Olefination* (pp. 1–17). Wiley-VCH Verlag GmbH and Co. Weinheim.

Egbe, D. A. M. et al., (2011). Alkoxy-substituted poly(arylene-ethynylene)-alt-poly(arylene-vinylene)s: synthesis, electroluminescence and photovoltaic applications. *J. Mater. Chem.*, *21*, 1338.

Espinet, P., & Echavarren, A. M., (2004). The mechanisms of the Stille reaction. *Angew. Chem. Int. Ed.*, *43*(36), 4704–4734.

Farina, V. (1996). New perspectives in the cross-coupling reactions of organostannanes. *Pure Appl. Chem. 68*, 73–78.

Farina, V., et al., (1998). The Stille reaction. In: *Book Review: Organic Reactions*. John Wiley & Sons, Inc.; New York.

Fichou, D., (1999). *Handbook of Oligo-and Polythiophenes*. Wiley-VCH. Weinheim.

Fleming, R., et al., (2014). Synthesis and thermal behavior of polyacrylonitrile/vinylidene chloride copolymer. *Polímeros*, *24*(3), 259–268.

Friend, R. H., et al., (1999). Electroluminescence in conjugated polymers. *Nature*, *397*, 121–128.

Gaboriaud, F., & Vantelon, J. P., (1982). Mechanism of thermal degradation of polyurethane-based on MDI and propoxylatedtrimethylol propane. *J. Polym. Sci. Part A-Polym. Chem.*, *20*(8), 2063–2071.

Gao, X., et al., (2014). Ultrasound-assisted Suzuki coupling reaction for rapid synthesis of polydihexylfluorene. *Polymer*, *55*(14), 3083–3086.

Genet, P. J., & Savignac, M. (1999). Recent developments of palladium(0) catalyzed reactions in aqueous medium. *J. Organomet. Chem.*, 576(1–2), 305–317.

Gjiri, E., (2020). *Knoevenagel Condensation*. Available at: http://www.name-reaction. com/knoevenagel-condensation (accessed on 17 June 2020).

He, J., et al., (2007). The effects of different side groups on the properties of polythiophene. *J. Macromol. Sci. Part A Pure Appl. Chem.*, *44*, 989–993.

Heeger, A. J., (2001). Semiconducting and metallic polymers: The fourth generation of polymeric materials (Nobel Lecture). *J. Phys. Chem. B.*, *105*(36), 8475–8491.

Horner, L., et al., (1958). Phosphororganische verbindungen, XII. Phosphinoxyde als olefinierungsreagenzien. *Chem. Ber.*, *91*, 61–63.

Horner, L., et al., (1959). Phosphinoxyde als olefinierungsreagenzien. *Chem. Ber.*, *92*(10), 2499–2505.

Hu, X., & Xu, L., (2000). Structure and properties of 3-alkoxy substituted polythiophene synthesized at low temperature. *Polymer*, *41*, 9147–9154.

Iovu, M. C., et al., (2005). Experimental evidence for the quasi-"living" nature of the Grignard metathesis method for the synthesis of regioregular poly(3-alkylthiophenes). *Macromolecules*, *38*(21), 8649–8656.

Jaballah, N., et al. (2011). Blue-luminescent poly(p-phenylenevinylene) derivatives: synthesis and effect of side-group size on the optical properties. *Eur. Polym. J.*, *47*(1), 78–87.

Kanatzidis, M. G., (1990). Conductive polymers. *Chem. Eng. News*, *68*(49), 36–50.

Kappe, C. O., & Dallinger, D., (2009). Controlled microwave heating in modern organic synthesis: Highlights from the 2004–2008 literature. *Mol. Divers*, *13*, 71–193.

Karpagam, S., et al. (2011). Applications of Wittig reactions in dibenzo 18-crown-6-ether substituted phenylenevinylene oligomer – synthesis, photo luminescent, and dielectric properties. *J. Appl. Polym. Sci.*, *120*(2), 960–967.

Knoevenagel, E., (1898). Condensation von malonsaure mit aromatischen aldehyden durch ammoniak und Amine. *Berichte der Dtsch. Chem. Gesellschaft.*, *31*(3), 2596–2619.

Koeckelberghs, G., et al., (2006). Regioregularity in poly(3-alkoxythiophene)s: Effects on the Faraday rotation and polymerization mechanism. *Macromol. Rapid Commun.*, *27*(22), 1920–1925.

Kosugi, M., & Fugami, K. (2002). A historical note of Stille reaction. *J. Organomet. Chem.*, *653*(1–2), 50–53.

Kotha, S., et al., (2002). Recent applications of the Suzuki-Miyaura cross-coupling reaction in organic synthesis. *Tetrahedron*, *58*, 9633–9695.

Lampman, G. M., Koops, R. W., & Olden, C. C., (1985). Phosphorus, and sulfur ylides formation; preparation of 1-benzoyl-2-phenylcyclopropane and 1,4-diphenyl-1,2-butadiene by phase transfer catalysis. *J. Chem. Educ.*, *62*(3), 267–268.

Larhed, M., et al., (1996). Rapid microwave-assisted Suzuki coupling on solid-phase. *Tetrahedron Lett.*, *37*(45), 8219–8222.

Laughlin, B. J., & Smith, R. C., (2010). Gilch and Horner-Wittig routes to poly(p-phenylenevinylene) derivatives incorporating monoalkyl defect-free 9,9-dialkyl-1,4-fluorenylene units. *Macromolecules*, *43*(8), 3744–3749.

Leclerc, M., & Faid, K., (1997). Electrical and optical properties of processable polythiophene derivatives: Structure-property relationships. *Advanced Materials*, *9*(14), 1087–1094.

Lee, L. H., (1964). Mechanisms of thermal degradation of phenolic condensation polymers. I. Studies on the thermal stability of polycarbonate. *J. Polym. Sci. Part A*, *2*(6), 2859–2873.

Lère-Porte, J. P., et al., (1999). Synthesis of regioregular poly(alkylbithiophene)s by an organometallic cross-coupling reaction. *Synth. Met.*, *101*(1–3), 588–589.

Levesque, I., & Leclerc, M., (1995). Ionochromic effects in regioregular ether-substituted polythiophenes. *J. Chem. Soc., Chem. Commun.*, 2293–2294.

Li, X. C., et al., (1999). Synthesis, properties, and application of new luminescent polymers with both hole and electron injection abilities for light-emitting devices. *Chem. Mater.*, *11*(6), 1568–1575.

Li, Y., et al., (1999). Electrochemical properties of luminescent polymers and polymer light-emitting electrochemical cells. *Synth. Met.*, *99*, 243–248.

Lux, A., et al., (1997). New CF3-substituted PPV-type oligomers and polymers for use as hole blocking layers in LEDs. *Synth. Met.*, *84*(1–3), 293–294.

Ma, H., et al., (2002). Highly efficient and thermally stable electro-optical dendrimers for photonics. *Adv. Funct. Mater.*, *12*(9), 565–574.

MacDiarmid, A. G., & Epstein, A. J., (1991). "Synthetic metals": A novel role for organic polymers (Nobel Lecture). *Makromol. Chem., Macromol. Symp.*, *51*, 11–28.

Maior, R. M. S., et al., (1990). Synthesis and characterization of two regiochemically defined Poly(dialkylbithiophenes): A comparative study. *Macromolecules*, *23*(5), 1268–1279.

Mark, H. F., (2011). Addition polymerization. In: *Encyclopedia of Polymer Science and Technology* (Vol. 1, pp. 444, 445). John Wiley & Sons, New Jersey.

Maryanoff, B. E., & Reitz, A. B., (1989). The Wittig olefination reaction and modifications involving phosphoryl-stabilized carbanions. Stereochemistry, mechanism, and selected synthetic aspects. *Chem. Rev.*, *89*(4), 863–927.

McCullough, R. D., & Lowe, R. D., (1992). Enhanced electrical conductivity in regioselectively synthesized poly(3-alkylthiophenes). *J. Chem. Soc., Chem. Commun.,* 70–72.

McCullough, R. D., (1998). The chemistry of conducting polythiophenes. *Adv. Mater., 10*(2), 93–116.

McCullough, R. D., et al., (1997). Self-assembly and disassembly of regioregular, water soluble polythiophenes: Chemoselective ionchromatic sensing in water. *J. Am. Chem. Soc., 119*(3), 633–634.

Mehtaa, V. P., & Eycken, E. V. V., (2011). Microwave-assisted C-C bond-forming cross-coupling reactions: An overview. *Chem. Soc. Rev., 40*(10), 4925–4936.

Melucci, M., et al., (2002). Microwave-assisted synthesis of thiophene oligomers via Suzuki coupling. *J. Org. Chem., 67*(25), 8877–8884.

Metzler, L., et al. (2015). High molecular weight mechanochromic spiropyran main chain copolymers via reproducible microwave-assisted Suzuki polycondensation. *Poly. Chem. 6*(19), 3694–3707.

Milstein, D., & Stille, J. K., (1978). A general, selective, and facile method for ketone synthesis from acid chlorides and organotin compounds catalyzed by palladium. *J. Am. Chem. Soc., 100*(11), 3636–3638.

Milstein, D., & Stille, J. K., (1979). Palladium-catalyzed coupling of tetraorganotin compounds with aryl and benzyl halides. Synthetic utility and mechanism. *J. Am. Chem. Soc., 101*(17), 4992–4998.

Mitchell, T. N., (1986). Transition-Metal catalysis on organotin chemistry. *J. Organomet. Chem., 304*(1–2), 1–16.

Mitchell, T. N., (1992). Palladium-catalysed reactions of organotin compounds. *Synthesis, 9,* 803–815.

Miyaura, et al., (1981). The palladium-catalyzed cross-coupling reaction of phenylboronic acid with haloarenes in the presence of bases. *Synth. Commun., 11*(7), 513–519.

Molina, R., et al., (2009). Progress in the synthesis of poly(2,7-fluorene-*alt*-1,4-phenylene), PFP, via Suzuki coupling. *Macromolecules, 42*(15), 5471–5477.

Moratti, S. C., et al., (1995). High electron-affinity polymers for LEDs. *Synth. Met., 71*(1–3), 2117–2120.

Morin, P. O., et al., (2016). Realizing the full potential of conjugated polymers: Innovation in polymer synthesis. *Mater. Horiz., 3,* 11–20.

Müllen, K., Reynolds, J. R., & Masuda, T., (2014). *Conjugated Polymers: A Practical Guide to Synthesis.* RSC Publishing. Cambridge.

Nehls, B. S., et al. (2004). Semiconducting polymers via microwave-assisted Suzuki and Stille cross-coupling reactions. *Adv. Func. Mater., 14*(4), 352–356.

Neilson, A. R., et al., (2007). Mixed chromophore perfluorocyclobutyl (PFCB) copolymers for tailored light emission. *Macromolecules, 40*(26), 9378–9383.

Ohshita, J., et al., (2009). Hole-injection properties of annealed polythiophene films to replace PEDOT-PSS in multilayered OLED systems. *Synth. Met., 159,* 214–217.

Pascariu, A., et al., (2003). Wittig and Wittig-Horner reactions under phase transfer catalysis conditions. *Cent. Eur. J. Chem., 1*(4), 491–542.

Pfeiffer, S., & Hörhold, H.-H., (1999). Investigation of poly(arylene vinylene)s, 41a. Synthesis of soluble dialkoxy-substituted poly(phenylene alkenylidene)s by applying the Horner-reaction for condensation polymerization. *Macromol. Chem. Phys.*, *200*, 1870-1878.

Qiu, Y., et al., (2016). Nickel-catalyzed Suzuki polymerization for controlled of ester-functionalized conjugated polymers. *Macromolecules*, *49*(13), 4757–4762.

Restrepo-Cossio, A. A., et al., (1997). Theoretical study of the mechanism of the Wittig reaction: Ab initio and mndo-pm3 treatment of the reaction of unstabilized, semi stabilized and stabilized ylides with acetaldehyde. *Heteroat. Chem.*, *8*(6), 557–569.

Sadhanala, A., et al., (2015). Electroluminescence from organometallic lead halide perovskite-conjugated polymer diodes. *Adv. Electron. Mater.*, *1*(3), 1500008.

Schlüter, A. D., (2001). The 10th anniversary of Suzuki polycondensation (SPC). *J. Polym. Sci. Part A Polym. Chem.*, *39*(10), 1533–1556.

Schlüter, A. D., et al., (2012). *Synthesis of Polymers: New Structures and Methods* (1st edn.). Wiley-VCH, Weinheim.

Seeboth, A., et al., (2014). Thermochromic polymers-function by design. *Chem. Rev.*, *114*(5), 3037–3068.

Sharma, N., et al., (2018). *Microwave-Assisted Organic Synthesis: Overview of Recent Applications*. (pp. 441–468). In: *Green Techniques for Organic Synthesis and Medicinal Chemistry*. Second Edition, John Wiley & Sons, Inc.; New York.

Sheina, E. E., et al., (2005). Highly conductive, regioregular alkoxy-functionalized polythiophenes: A new class of stable, low band gap materials. *Chem. Mater.*, *17*(13), 3317–3319.

Shinde, K. N., et al., (2012). Basic mechanisms of photoluminescence. In: *Phosphate Phosphors for Solid-State Lighting* (Vol. 174, pp. 41–59). Springer Series in Materials Science, Berlin.

Shirakawa, H., (2001). The discovery of polyacetylene film-the dawning of an era of conducting polymers (Nobel Lecture). *Curr. Appl. Phys.*, *1*, 281–286.

Stefan, M. C., et al., (2009). Grignard metathesis method (GRIM): Toward a universal method for the synthesis of conjugated polymers. *Macromolecules*, *42*(1), 30–32.

Stein, P. C., et al., (1995). NMR study of the structural defects in poly(3akylthiophene)s: Influence of the polymerization method. *Synth. Met.*, *69*, 305–306.

Still, W. C., & Gennari, C. (1983). Direct synthesis of Z-unsaturated esters. A useful modification of the Horner-Emmons olefination. *Tetrahedron Lett.* 24(41), 4405–4408.

Stille, J. K., (1986). The palladium-catalyzed cross-coupling reactions of organotin reagents with organic electrophiles. *Angew. Chem. Int. Ed. Engl.*, *25*(6), 508–524.

Sugimoto, R., et al., (1985). Preparation and property of polytellurophene and polyselenophene. *Jpn. J. Appl. Phys.*, *24*(6), L425–L427.

Suzuki, Y., et al., (2007). Synthesis of regioregular poly(*p*-phenylenevinylene)s by Horner reaction and their regioregularity characterization. *Macromolecules*, *40*, 6521–6528.

Tajima, K., et al., (2008). Polymer photovoltaic devices using fully regioregular poly[(2-methoxy-5-(3',7'-dimethyloctyloxy))-1,4-phenylenevinylene]. *J. Phys. Chem. C.*, *112*(23), 8507–8510.

Tamao, K., et al., (1982). Nickel-phosphine complex-catalyzed Grignard coupling-II. Grignard coupling of heterocyclic compounds. *Tetrahedron*, *38*(22), 3347–3354.

Tierney, S., et al., (2005). Microwave-assisted synthesis of polythiophenes via the stille coupling. *Synth. Met.*, *148*(2), 195–198.

Tietze, L. F., & Beifuss, U., (1992). The Knoevenagel Reaction. In: *Comprehensive Organic Synthesis* (Vol. 2, pp. 341–394). Pergamon Press; Oxford.

Torres, B. B. M., & Balogh, D. T., (2012). Regioregular improvement on the oxidative polymerization of poly-3-octylthiophenes by slow addition of oxidant at low temperature. *J. Appl. Polym. Sci.*, *124*, 3222–3228.

Thompson, S. K., & Heathcock. C. H., (1990). Effect of cation, temperature, and solvent on the stereoselectivity of the Horner-Emmons reaction of trimethyl phosphonoacetate with aldehydes. *J. Org. Chem.*, *55*, 3386–3388.

Truong, M. A., et al., (2015). Synthesis, characterization, and application to polymer solar cells of polythiophene derivatives with ester-or ketone-substituted phenyl side groups. *J. Polym. Sci. Part A – Polym. Chem.*, *53*(7), 875–887.

Varnado, C. D., & Bielawski, C. W., (2012). Condensation polymers via metal-catalyzed coupling reactions. *Polym. Sci. A Compr. Ref.*, *5*, 175–194.

Vázquez-Guilló, R., et al. (2018). Advantageous microwave-assisted Suzuki polycondensation for the synthesis of aniline-fluorene alternate copolymers as molecular model with solvent sensing properties. *Polymers*, *10*(2), 215–230.

Vedejs, E., & Peterson, M. J., (1994). In: *Topics in Stereochemistry*. Vol. 21, 1–158. John Wiley & Sons, Inc.; New York.

Vedejs, E., & Peterson, M. J., (1996). In: *Advances in Carbanion Chemistry*. Vol. 2, 1–85. JAI Press; New York.

Wadsworth Jr., W. S., & Emmons, W. D., (1961). The utility of phosphonate carbanions in olen synthesis. *J. Am. Chem. Soc.*, *83*, 1733–1738.

Wang, F., et al., (2008). Poly(p-phenylene vinylene) derivatives with different contents of cis-olefins and their effect on the optical properties. *Macromol. Chem. Phys.*, *209*, 1381–1388.

Wittig, G., & Geissler G., (1953). Zur reaktionsweise des pentaphenyl-phosphors und einiger derivate. *Liebigs Ann.*, *580*, 44–57.

Wu, J., et al., (2015). Suzuki polycondensation and post-polymerization modification toward electro-optic perfluorocyclobutyl (PFCB) aryl ether polymers: Synthesis and characterization. *J. Fluor. Chem.*, *180*, 227–233.

Xu, B., et al., (2014). Fine-tuning of polymer properties by incorporating strongly electron-donating 3-hexyloxythiophene units into random and semi-random copolymers. *Macromolecules*, *47*(15), 5029–5039.

Yang, Z., & Geise, H. J. (1992). Preparation and electrical conductivity of blends consisting of modified Wittig poly(para-phenylene vinylene), iodine and polystyrene, polymethyl methacrylate or polycarbonate. *Synth. Met.*, *47*(1), 105–109.

Yokoyama, A., & Yokozawa, T., (2009). Chain-growth condensation polymerization for the synthesis of well-defined condensation polymers and π-conjugated polymers. *Chem. Rev.*, *109*(11), 5595–5619.

Yokoyama, A., et al., (2007). Chain-growth polymerization for the synthesis of polyfluorene via Suzuki-Miyaura coupling reaction from an externally added initiator unit. *J. Am. Chem. Soc.*, *129*(23), 7236–7237.

Yoshino, K., et al., (1984). Preparation and properties of conducting heterocyclic polymer films by chemical method. *Jpn. J. Appl. Phys.*, *23*(12), L899–L900.

Zerong, W., (2010). Knoevenagel condensation. In: *Comprehensive Organic Name Reactions and Reagents* (pp. 1621–1626). John Wiley & Sons, Inc.

Zhang, H. H., et al., (2016). t-Bu3P-coordinate 2-phenylaniline-based palladacycle complex/ArBr as robust initiators for controlled Pd(0)/t-Bu3P-catalyzed Suzuki cross-coupling polymerization of AB-type monomers. *ACS Macro Lett.*, *5*(6), 656–660.

Zhang, W., et al., (2013). Microwave-assisted Suzuki coupling reaction for rapid synthesis of conjugated polymer poly(9,9-dihexylfluorene)s as an example. *J. Polym. Sci. Part A-Polym. Chem.*, *51*(9), 1950–1955.

Ziabicki, A., & Kedzierska, K., (1962). Studies on the orientation phenomena by fiber formation from polymer melts: III. Effect of structure on orientation, condensation polymers. *J. Appl. Polym. Sci.*, *6*(19), 111–119.

CHAPTER 2

Photophysics: Basic Concepts

JOSÉ CARLOS GERMINO and LUÍS GUSTAVO TEIXEIRA ALVES DUARTE

Chemistry Institute, University of Campinas, Campinas, Brazil

2.1 MOTIVATION

The consequent physicochemical dynamics from light absorption on the ultraviolet and visible regions (UV/Vis) by molecular systems are the object of study to many scientific fields. The development of such systems has countless applications including the design of optoelectronic devices (Guo et al., 2013; Reiss, 2011) such as organic light-emitting diodes (OLEDs) (Orgiu and Samori, 2014; Reineke et al., 2013; Uoyama et al., 2012), solar cells (Kang et al., 2016; Mazzio and Luscombe, 2015; Facchetti, 2011), sensors (Sonawane and Asha, 2016; Singh et al., 2016; Baeg et al., 2013; Chen et al., 2011) and imaging (Lerch et al., 2016; Reineke et al., 2013; Zheng et al., 2013; Ito et al., 2010). In this sense, the purpose of this chapter is to present an overview of basic concepts related to the photophysics of organic aromatic molecules. These ideas will be built starting from small molecules until macromolecules, specifically semiconducting polymers (SemPolys), focusing on their applications.

One of our objectives is to describe how the luminescence phenomenon occurs and how it impacts in OLEDs performance, focusing mostly on energy transfer processes, particularly the mechanism described by Theodore Förster (1959). Also, with this chapter, the reader can feel comfortable on going through the book chapters minimizing the need for other texts, which may be helpful to students and new researchers to understand the ideas concerning all books.

Thus, the most straightforward way, to begin with, is describing the Jablonski Diagram. From the diagram, one can visualize and assign not only electronic transitions in a qualitative way but also quantitative information that derives from thermodynamics and kinetics. Besides, it is necessary to stress that the examples used to explain every step where taken from the recent literature, so the notions do not turn to abstract.

2.2 PHOTOPHYSICAL PROCESSES

An organic aromatic molecular system returns to the ground electronic state, after absorption of electromagnetic radiation, through dissipation of the acquired energy (Valeur, 2002). This return is not always simple; it can involve different deactivation pathways (Rohatgi, 1978), divided into photophysical and photochemical processes. A photochemical process differs from the photophysical once it promotes molecular alterations due to a chemical reaction triggered by light absorption.

Condensed matter presents several deactivation paths from electronically excited states; some of them constitute intrinsic properties (unimolecular processes); others are derived from external perturbations (bimolecular processes) (Birks, 1970). All of them respect the basic principles regarding the Jablonski diagram (Figure 2.1).

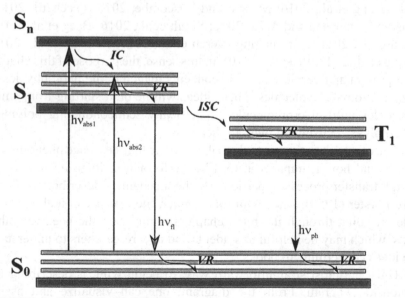

FIGURE 2.1   The Jablonski diagram.

The diagram contains all relative energies of typical photophysical steps. Starting from ground state electronic configuration, an organic aromatic molecular system undergoes a transition to excited state electronic configuration due to momentum transfer from light, which must contain specific energy that resonates with the electric density of the molecule. This event characterizes an electronic transition and its probability follows the Franck-Condon Principle, mathematically expressed by:

$$P = |\langle \psi_i | M | \psi_j \rangle|^2 |\langle X_a | X_b \rangle|^2 \qquad (2.1)$$

where, ψ_i and ψ_j are the electronic wavefunctions of two states, with χ_a and χ_b being the vibrational wavefunctions. M is the transition dipole moment operator, and acts just on the electronic package (Valeur, 2002). Rotational energies can be neglected since they do not imply significant energy changes. In fact, Eq. (2.1) is responsible for spectral bands intensities because it allows the verification of the electronic transition probability related with wavefunctions overlap. Thus, the physical meaning of a spectral band is a distribution of states (McQuarrie and Simon, 1997).

Once in the excited electronic state, a generic molecular system is ready to dissipate the accumulated energy from excitation, this happening by the emission of heat or radiation, where the first one is named vibrational relaxation (VR). VR comes from the deactivation of vibrational levels on the same electronic level. Internal conversion (IC) comes from different electronic levels, hence the energy difference from $h\nu_{abs1}$ to $h\nu_{abs2}$ in Figure 2.1. Those are *non-radiative* paths of excited states deactivation, which are typically evaluated by the Stokes Shift (SS): the difference between the absorption and emission energies.

The second one receives other two distinct names, both assigned as Luminescence; radiative paths to deactivate an excited state and start solely from the first excited electronic state (S_1) for further electronic states are very close in energy, disabled by IC. These paths depend on molecular orbital symmetry between the ground and excited electronic states. Besides, if the electronic transition does not promote spin change, after VR and/or IC, the return to S_0 radiatively is named Fluorescence. On another hand, the electronic transition from S_0 to S_1 may result in the spin change (intersystem crossing (ISC)), then the Triplet state (T_1), an intermediate excited state, is originated, also deactivating radiatively by a process named phosphorescence (Lakowicz, 2006).

2.3 EXCITED STATE FORMATION

The Franck-Condon principle stands for the electronic transition probability and is responsible for the registered signal of an electronic transition after excitation. The formation of an excited state is a consequence of the population of different states with different energies. In other words, a spectral band reflects a probability distribution that will be centered at the most effective transition, which will always be of high occurrence because of molecular orbitals wavefunction overlap.

A way to account for this transition probability is the analysis of the spectral band intensity on the electronic absorption spectrum by the extinction coefficient "ε." This greatness is a physical parameter that comes from the sample effective irradiated cross-section "σ," following the expression:

$$I = I_i e^{-\sigma n d} \tag{2.2}$$

where, I_i is the initial intensity of the radiation and n the number of molecules per cm^3 inside of a region with length d.

In the most useful way, especially in chemistry, it is more interesting to write Eq. (2.2) in terms of molar concentration "[Y]." So, using the Avogadro number N:

$$n = N[Y]10^{-3} \tag{2.3}$$

The combination of Eqs. (2.2) and (2.3) gives:

$$I = I_i 10^{-\varepsilon[Y]d} \tag{2.4}$$

Above is the Lambert-Beer Law, which contains the relation between absorbed radiation and the concentration of the sample. Commonly, the electronic absorption spectrum is plotted in terms of ε or $log\varepsilon$.

The extinction coefficient value, or molar absorptivity coefficient, allows the description of the absorption spectrum by the oscillator strength f, considering a molecule as an oscillating dipole that is altered by irradiation:

$$f = 2303\frac{mc^2}{N\pi e^2 n}\int_0^\infty \varepsilon\left(\tilde{v}\right)d\tilde{v} \tag{2.5}$$

where, f is a normalized dimensionless quantity, with a maximum that matches with the one from absorption maximum, m and e are the mass and charge of the electron, n the refractive index of the medium and c is the speed of the light (Valeur, 2002; Birks, 1970).

Due to the high molar mass of SemPolys, their molar absorptivity coefficient cannot be evaluated through the simple use of molar concentration on Eq. (2.4) because it produces an overestimated value. To solve this issue, typically they are roughly estimated using the molar mass of the monomeric unit (Vezie et al., 2016).

In general, when the system possesses alternated single and double/triple bonds, SemPolys presents a π-conjugated framework that is directly linked to its electronic transition, so the highest occupied molecular orbital (HOMO) is a π-orbital, and the lowest occupied molecular orbital (LUMO) is a π^*-orbital, both coming from the overlap of a p_y or a p_z atomic orbitals- an electronic transition allowed by symmetry. The resulting energy gap created from this HOMO-LUMO electronic transition (a π-π^* transition) varies from 1.5 eV to 3.0 eV ($\approx$ 827 nm to $\approx$ 414 nm), characterizing a semiconductor material.

The first studies of SemPolys were performed by the same researchers that were initially studying inorganic semiconductors. Thereby, initially, the same approach of the Band Model for solids was used to explain SemPolys transitions. The difference was in the fact that valence and conduction bands were approximated by the sum of molecular ligand π and anti-ligand π^* orbitals. However, this model is not realistic once, for instance, does not account for the existence of electron-electron, electron-phonon coupling, or even triplet states. Furthermore, SemPolys present an enormous spectral variety, from symmetric and fine bands to dimensionless vibronic structures. In this sense, as most of SemPolys technological applications come from systems in the solid-state, crystal orientation may offer spectral information relative to molecules on the gas phase, contradicting expected results. Thus, to explain the spectral features of such organic semiconductors, the Exciton Model was developed.

An exciton is defined as a quasi-particle chargeless and diffuser, a bound state of an electron-hole pair dependent on photoexcitation or charge-carriers injection (which is the case of optoelectronic devices), with the pair interacting by Coulombic potential. It also has binding energy, before electron dissociation and diffusion, so, to SemPolys, its energy value relative to ground state will be less than the bandgap and now maybe originated from a singlet, triplet, or trap states (the last one related to polymer chain disorder or extension).

2.4 ANALYTICAL SOLUTIONS TO RECOMBINATION RATES

From Einstein's equations to the probability of an electronic transition and deactivation by spontaneous or induced emission (unimolecular process). In 1962, Strickler and Berg derived the relationship between absorption intensity in terms of ε, and the average residence time of the excited state, i.e., the emission lifetime or even the recombination rate of excited electronic state (Strickler and Berg, 1962). Considering as boundary conditions an intense peak of absorption ($\varepsilon > 8000\ L\ mol^{-1}\ cm^{-1}$) that can be approximated to a Gaussian function and that the molecule does not suffer many variations on its electronic distribution when excited, the Strickler-Berg Relation (SB Relation) is expressed by the following equation:

$$\frac{1}{\tau_0} = A_{u \to l} = 2.88 \times 10^{-9}\, n^2 \left\langle \upsilon_f^{-3} \right\rangle_{A_{u \to l}}^{-1} \frac{g_u}{g_l} \int \varepsilon\, dln\upsilon \qquad (2.6)$$

where, τ_0 is the emission lifetime, $A_{u \to l}$ is the Einstein coefficient to spontaneous emission from an upper level u to a lower level l, g_u and g_l are the degeneracies of both states, $\tilde{\upsilon}$ represents the frequency transition, and $\left\langle \tilde{\upsilon}^{-3} f \right\rangle A_{u \to l}$ stands for:

$$\left\langle \tilde{\upsilon}_f^{-3} \right\rangle A_{u \to l} = \frac{\int I\!\left(\tilde{\upsilon} \right) d\tilde{\upsilon}}{\int \tilde{\upsilon}^{-3} I(\tilde{\upsilon}) d\tilde{\upsilon}} \qquad (2.7)$$

That is a ratio of integrals along the whole emission spectrum. To many authors, τ_0 is the "natural emission lifetime," once it accounts for the pure transition of two consecutive levels. Thus, it might differ from the measured emission lifetime because of the other photophysical phenomena described by the Jablonski Diagram (Figure 2.1).

Besides the SB Relation, many researchers developed methods to obtain the emission lifetime. Among them, another classical equation was evaluated by van Roosbroeck and Schokley in 1954 to inorganic semiconductors, using nothing, but the absorption spectrum:

$$\frac{1}{\tau_0} = \frac{n_0 + p_0}{n_0\,p_0}\frac{8_\pi}{c^2}\int\frac{a_v\,n_v^2\,v^2}{hv}dv$$
$$e^{KT-1}$$

$$(2.8)$$

From Eq. (2.8), n_0 and p_0 are the concentration of charger carriers at the conduction and valence band, respectively. a_v and n_v^2 are the extinction coefficient and refractive index at frequency v. In addition, the Roosbroeck-Schokley equation considers the thermal energy kT, as a Boltzman Distribution on the description, being k the Boltzman constant and T the temperature.

Later in 1991, to universalize the ability of emission lifetime prediction, J. R. Bolton and Mary D. Archer derived a set of methods with distinct boundary conditions to calculate τ_0, which differs small chromophores from semiconductors. Their equations summarize recombination lifetimes by means of their rates k_{rl} to small chromophores τ_0^{sc} (Eq. (2.2)) or semiconductors τ_0^{se} (Eq. (2.10)) as follows:

$$\tau_0^{sc} = \frac{1}{k_{r1}} \tag{2.9}$$

and

$$\tau_0^{se} = \frac{1}{(n_0 + p_0)k_{r2}} \tag{2.10}$$

In their work, one of the methods can be applied only to systems in which electronic absorption and fluorescence spectra possess an overlap region. However, small organic aromatic molecules not always have an overlap region available due to low extinction coefficients on the region and large SS. Nevertheless, it depends on the optical band gap energy U_g, or the energy of pure electronic levels transition "0–0," again not counting vibrational or rotational energy levels in between.

Under these circumstances, considering the overlap region $f(v)$, with corresponding extinction coefficient and refractive index a_v and n_v^2, their so-called "Method 1" follows the next two set of equations for molecules (Eq. (2.11)) and inorganic semiconductors (Eq. (2.12)):

$$\ln\left[\frac{f(v)}{\alpha_v n_v^2 v^2}\right] = \ln\left[\frac{8\pi \times 0.2303\tau_0^{sc}}{Nc^2}\frac{q_l}{q_u}\right] + \frac{U_g - hv}{kT} \tag{2.11}$$

and

$$\ln\left[\frac{f(v)}{\alpha_v n_v^2 v^2}\right] = \ln\left[\frac{8\pi \times (n_0 + p_0)\tau_0^{se}}{N_c N_v c^2}\frac{q_l}{q_u}\right] + \frac{U_g - hv}{kT} \tag{2.12}$$

Once more, likewise Roosbroeck-Schokley equation, thermal energy is considered. Besides, the partition function to both states (u and l) become relevant, despite not always known and being approximate to the unity.

Now, Eqs. (2.11) and (2.12) are straight lines and the emission lifetimes are inside the linear coefficient. Bolton and Archer applied the equations to the dye Rhodamine 6G. The spectrum and plot of the equation are depicted in Figure 2.2.

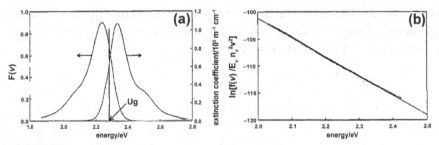

FIGURE 2.2   (a) Electronic absorption and emission spectra to Rhodamine 6G in ethanol at 296 K, (b) resulting plot from Eq. (2.11) referent to the dye.

Source: Reprinted/adapted with permission from Bolton and Archer, 1991.

Rhodamine 6G has an experimental emission lifetime of . The recombination time estimated by SB Relation gave $\tau_0 = 4.003$ ns and the use of Eq. (2.11) offered .

2.5   FLUORESCENCE QUANTUM YIELD

Concerning the description of the unimolecular processes of excited electronic states deactivation, it can be cited that another photophysical parameter of great importance is the fluorescence quantum yield Φ, the ratio between the number of emitted and absorbed photons. Considering a generic system ^1Y on the excited state, with k_F as the emission rate and k_{nr} the sum of all non-radiative deactivation rates of the specie $^1Y*$ (Eq. (2.13)) assigned by the Jablonski Diagram, one can write its formation rate through the following kinetic equation:

$$\frac{d\left[^1Y^*\right]}{dt} = I_0 - \left(k_F + k_{nr}\right)\left[^1Y^*\right] \tag{2.13}$$

and

$$k_{nr} = k_{IC} + k_{ISC} \tag{2.14}$$

where, $\dfrac{d\left[^1Y^*\right]}{dt}$ is the formation rate of $^1Y*$; I_0 is the intensity of the photons absorbed by ^1Y; k_{IC} and k_{ISC} are the IC and the inter-system crossing rates constants, respectively.

To photostationary conditions, $\dfrac{d\left[^1Y^*\right]}{dt} = 0$ so:

$$I_0 = \left(k_F + k_{nr}\right)\left[^1Y^*\right] \tag{2.15}$$

Taking Eq. (2.15), the fluorescence quantum yield of $^1Y*$ is described by:

$$\Phi = \int_0^\infty F(v)\,dv = \frac{\#emitted photons}{\#absorbed photons} = \frac{k_F\left[^1Y^*\right]}{I_0} = \frac{k_F}{k_F + k_{nr}} \tag{2.16}$$

where, $F(v)$ is the normalized fluorescence spectrum for $^1Y*$.

Eq. (2.16) represents the ratio between radiative and the sum of radiative and non-radiative processes that might take place with $^1Y*^1$.

The same idea is used to photodynamic (or transient) conditions, depending on the emission lifetime. Considering the process after formation of $^1Y*$, and that the intensity of transient emission $I(t)$ is proportional to $^1Y*$ concentration ($I(t) \propto [^1Y*]$), the deactivation rate of $^1Y*$ becomes:

$$\frac{d\left[^1Y^*\right]}{dt} = \frac{dI(t)}{dt} = -\left(k_{FY} + k_{nrY}\right)I_Y \tag{2.17}$$

and

$$\frac{dI(t)}{I} = -\left(k_{FY} + k_{nr}\right)dt \tag{2.18}$$

Integrating Eq. (2.18) and applying the exponential function on both sides gives:

$$I(t) = I_0 e - (k_{FY} + k_{nr})^t \tag{2.19}$$

From Eq. (2.9) to photodynamic conditions, it's known that $k_{FY} + k_{nrY} = k_Y = \dfrac{1}{\tau_Y}$. Replacing this statement on Eq. (2.19) gives:

$$I(t) = I_0 e^{-\frac{t}{\tau_Y}} \tag{2.20}$$

Eq. (2.20) represents the fluorescence emission decay to the chromophore $^1Y^*$ as a monoexponentially decay, where, τ_Y is the fluorescence lifetime, t is the time and I_0 the normalized intensity at decay beginning (Guo et al., 2013; Lakowicz, 2006).

Now, Eq. (2.20) also stands for the emission profile of a single species on the excited electronic state. However, once in the excited state, several processes may occur, including conformational variations or even photoreactions. In the occurrence of such events, the same chromophore can present a multiexponential decay, so Eq. (2.20) changes a bit to:

$$I(t) = \sum_{i=1}^{n} B_i e^{-\frac{t}{\tau_i}} \tag{2.21}$$

where, τ_i is the fluorescence lifetime and B_i is the amplitude of each species on excited electronic state at $t = 0$ (relative percentage of each of the components).

Finally, $I(t)$ has a direct relation to Φ:

$$\Phi = \int_0^\infty I(t)\,dt = \frac{k_F}{k_F + k_{nr}} = \frac{\tau_Y}{\tau_{FY}} = \int_0^\infty F(v)\,dv \tag{2.22}$$

where, τ_{FY} is the lifetime of radiative processes only, which is the inverse of the fluorescence rate $\left(k_{FY} = \dfrac{1}{\tau_{FY}} \right)^{1}$ and the equivalent to τ_0, the "natural emission lifetime." Therefore, that is the origin of the difference from the emission lifetime calculated by Eqs. (2.6) and (2.11) to the experimentally determined.

2.6   KINETICS OF BIMOLECULAR PROCESSES

Bimolecular processes on the excited state are related to quenching processes. Fluorescence quenching depends on the presence of an energy quencher Q, which can be an impurity or something added on purpose. There are four basic types of fluorescence quenching: collisional, quenching due to excimer formation, inner filter effect, and quenching by energy transfer.

2.6.1   COLLISIONAL QUENCHING

In solution, quenching impurities with diffusion rates bigger than our generic chromophore ^{1}Y (which may be for instance dissolved oxygen, that is easier found in concentrations of 10^{-3} mol L^{-1} of magnitude in non-polar organic solvents), can collide with the excited species $^{1}Y^{*}$ (Figure 2.3), leading to its non-radiative deactivation and the consequent decrease of quantum yield.

FIGURE 2.3   Schematic representation of collisional quenching.

The collisional quenching mechanism can be described as:

$$\frac{d\left[^{1}Y^{*} \right]}{dt} = I_0 - \left(k_F + k_{nr} \right)\left[^{1}Y^{*} \right] - k_q \left[Q \right]\left[^{1}Y^{*} \right] \qquad (2.23)$$

where, k_q is the rate constant to the process and $[Q]$ is the concentration of the quencher. Considering $[Q]>>[^1Y^*]$, Eq. (2.23) becomes of pseudo-first-order with the product $k_q[Q]$ constant:

$$\frac{d\left[^1Y^*\right]}{dt} = I_0 - \left(k_F + k_{nr} + k_q[Q]\right)\left[^1Y^*\right] \tag{2.24}$$

Replacing Eq. (2.24) on Eq. (2.16) gives the quantum yield to photo-stationary conditions as follows:

$$\Phi = \frac{k_{FY}\left[^1Y^*\right]}{I_0} = \frac{k_F}{k_F + k_{nr} + k_q[Q]} \tag{2.25}$$

Thus, Eq. (2.25) demonstrates why the quantum yield lowers. Furthermore, assuming the condition of a photostationary state $((d[^1Y^*]/dt)=0)$, the ratio between Eqs. (2.15) and (2.25) gives rise to:

$$\frac{[^1Y^*]_0}{\left[^1Y^*\right]} = 1 + \frac{k_Q[Q]}{k_F + k_{nr}} \tag{2.26}$$

The analysis of Eq. (2.26) shows how the fluorescence intensity of the chromophore varies on the presence and absence of Q, once they are proportional to $[^1Y^*] \mid [^1Y^*]_0$, so:

$$\frac{F_0}{F} = 1 + \tau_Y k_Q[Q] \tag{2.27}$$

This last equation is known as the Stern-Volmer equation to fluorescence quenching, where, $\tau_Y k_Q$ is the Stern-Volmer constant K_{SV}.

2.6.2   QUENCHING DUE TO EXCIMER FORMATION

The increase of $^1Y^*$ concentration generally diminishes the fluorescence quantum yield; this may happen because of the formation of excimers. The higher is the concentration, the greater is the probability of two $^1Y^*$ molecules encounter each other and form a dimer on the excited state

($^1D*$). The mechanism is equivalent to collisional quenching, with the quantum yield following the next relation, with k_D being the excimer rate formation:

$$\Phi = \frac{k_F \left[{}^1Y^*\right]}{I_0} = \frac{k_F}{k_F + k_{nr} + k_D\left[{}^1Y\right]} \tag{2.28}$$

This quenching mechanism obeys a Stern-Volmer kinetics type too.

2.6.3 INNER FILTER EFFECT

Another quenching mechanism of fluorescence is by self-absorption, also named inner filter effect. This phenomenon is a consequence of spectral overlap from absorption and emission spectra. Generally, the overlap of the 0–0 transition on both spectra which is related to concentration increases. What happens is that part of the emitted energy from $^1Y*$ is reabsorbed by another molecule of ^1Y on the ground electronic state (Figure 2.4).

$$^1Y_A^* \xrightarrow{k_{FY}} {}^1Y_A + h\nu_{FY}$$

$$h\nu_{FY} + {}^1Y_B \longrightarrow {}^1Y_B^*$$

$$^1Y_B^* \xrightarrow{k_{FY}} {}^1Y_B + h\nu_{FY}$$

FIGURE 2.4 Mechanism to inner filter effect. "A" and "B" refer to different molecules of the same species.

This effect is neglected to chromophores on the first triplet excited state (T_1), once the absorption intensity of the electronic transition $S_0 \rightarrow T_1$ has a very low extinction coefficient.

Considering our system Y again, with a as the probability of an emitted photon to be reabsorbed by the same molecule on the ground state, and $1 - a$ as the probability of an emitted photon not to be reabsorbed, the expression to the quantum yield Φ to self-absorption is the power series:

$$\Phi = \phi(1-a)\left[1+a\phi+a^2\phi^2+...+a^n\phi^n\right] = \frac{\phi(1-a)}{1-aq_F} \qquad (2.29)$$

where, the power series successive terms represent the probabilities of reabsorption after 1, 2, 3, ..., n emissions. ϕ is the fluorescence quantum yield without the inner filter effect occurrence, and q_F the quantum efficiency (the ratio of emitted photons and the number of molecules on the excited state).

In this way, self-absorption diminishes the decay rate of $^1Y^*$ and enlarges the fluorescence lifetime. Also, their mathematical relations are:

$$\frac{d\left[^1Y^*\right]}{dt} = \left\{(1-a)k_{FY}+k_{nr}\right\}\left[^1Y^*\right] = \frac{\left[^1Y^*\right]}{\tau_{if}} \qquad (2.30)$$

and

$$\tau_{if} = \frac{1}{(1-a)k_{FY}+k_{nr}} = \frac{\tau_Y}{1-aq_{FY}} \qquad (2.31)$$

2.6.4   ENERGY TRANSFER

The understanding of energy transfer mechanisms is of great interest to many scientific fields. Among then, it can be cited: materials science [where energy and charge transfer processes play a relevant role in optoelectronic properties and efficiency of organic and polymeric light-emitting diodes (OLEDs and PLEDs) (Farinola and Ragni, 2011; Laquai et al., 2009)], photovoltaic cells [for which those events are used to explain their operation (Quites et al., 2014b)], biology (where there are many studies about non-radiative energy transfer in biochemical systems (Luhman and Holmes, 2011; Reiss et al., 2011)), chemistry [where through energy transfer one may obtain the actual distance between molecules and estimate its efficiency as a function of the distance and emission lifetimes from host to guest (Singh et al., 2016; Reiss et al., 2011)], and physics [which developed the main models to the comprehension of radiative and non-radiative energy, highlighting the Förster model (1946), described from quantum mechanical and classical physics (Singh et al., 2016; Reineke et al., 2013)].

2.6.4.1 RADIATIVE ENERGY TRANSFER

The radiative energy transfer process is also designated as the trivial model due to the simplicity of the physical process involved in the energy transfer from a donor (D*) to an acceptor (A*). Its mechanism is considered simple since the donor (a species on the excited state) deactivates radiatively through fluorescence ($^1D^*$) or phosphorescence ($^3D^*$), the acceptor (a species on the singlet ground, ^1A) absorbs this energy coming from the donor, goes to the excited state ($^1A^*$), and finally deceives too-radiatively or not.

Therefore, this process demands spectral overlap between the electronic emission spectrum of D* and the electronic absorption spectrum of ^1A. Figure 2.5 presents the proposed mechanism, which is relevant to diluted solutions or solid systems with two independent layers (Birks, 1970).

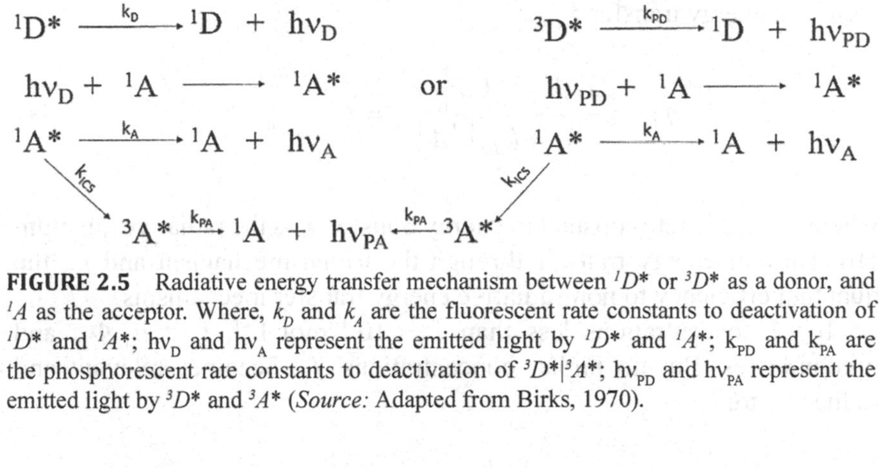

$$^1D^* \xrightarrow{k_D} {}^1D + hv_D \qquad\qquad {}^3D^* \xrightarrow{k_{PD}} {}^1D + hv_{PD}$$

$$hv_D + {}^1A \longrightarrow {}^1A^* \qquad or \qquad hv_{PD} + {}^1A \longrightarrow {}^1A^*$$

$$^1A^* \xrightarrow{k_A} {}^1A + hv_A \qquad\qquad {}^1A^* \xrightarrow{k_A} {}^1A + hv_A$$

$$^3A^* \xrightarrow{k_{PA}} {}^1A + hv_{PA} \xleftarrow{k_{PA}} {}^3A^*$$

FIGURE 2.5 Radiative energy transfer mechanism between $^1D^*$ or $^3D^*$ as a donor, and ^1A as the acceptor. Where, k_D and k_A are the fluorescent rate constants to deactivation of $^1D^*$ and $^1A^*$; hv_D and hv_A represent the emitted light by $^1D^*$ and $^1A^*$; k_{PD} and k_{PA} are the phosphorescent rate constants to deactivation of $^3D^*|^3A^*$; hv_{PD} and hv_{PA} represent the emitted light by $^3D^*$ and $^3A^*$ (*Source:* Adapted from Birks, 1970).

It is important to emphasize that for low-intensity transitions as $S_0 \rightarrow T_1$ and $T_0 \rightarrow T_1$, the probability a_{AD} of trivial energy transfer to triplet ground state ^3A is close to zero. At the same time, if the chromophore ^1A has a high ISC rate constant, the excited species $^1A^*$ will become $^3A^*$, being that the last one may be phosphorescent (Birks, 1970).

In a sense, trivial energy transfer is like the inner filter effect in solution and can be expressed by:

$$a_{AD} = \frac{1}{\Phi_D} \int_0^\infty F_D(\lambda)\left\{1 - 10^{-\varepsilon(\lambda)[^1A]x^{-1}}\right\} d\lambda \qquad (2.32)$$

and

$$a_{AD} \cong \frac{2{,}303\left[^1A\right]}{x\Phi_D}\int_0^\infty F_D(\lambda)\varepsilon_A(\lambda)d\lambda \qquad (2.33)$$

where, a_{AD} is the probability of trivial energy transfer occurrence between D and A, Φ_D is the fluorescence quantum yield of D in the absence of A. $[^1A]$ is the molar concentration of ^1A, x is the distance between $^1D*$ and ^1A.

The integral $\int_0^\infty F_D(\lambda)\varepsilon_A(\lambda)d\lambda$ represents the spectral overlap $J_{AD}(\lambda)$

between the emission spectrum of D* ($F_D(\lambda)$) and the absorption spectrum of ^1A ($\varepsilon_A(\lambda)$), as a function of the wavelength (Guo et al., 2013; Luhman and Holmes, 2011).

From the probability a_{AD}, it is possible to evaluate the quantum efficiency to energy transfer f_{AD}:

$$f_{AD} = \frac{a_{AD}k_{FD} + k_{AD}\left[^1A\right]}{k_D + k_{AD}\left[^1A\right]} = f_R + f_{NR} \qquad (2.34)$$

where, k_{AD} is the rate constant to energy transfer, f_R is the radiative quantum efficiency of energy transfer through the trivial mechanism and f_{NR} the quantum efficiency to non-radiative energy transfer mechanisms.

To ^1A concentrations less than 1×10^{-4} mol L^{-1}, $f_R = a_{AD}\Phi_D$, and predominates. But, to higher concentrations, f_{NR} becomes relevant and reduces f_R to:

$$f_R = \frac{a_{AD}k_{FD}}{k_{FD} + k_{nrD} + k_{AD}\left[^1A\right]} = a_{AD}\Phi_{AD} \qquad (2.35)$$

with Φ_{AD} being the fluorescence quantum yield of D in the presence of A.

Eq. (2.34) can be interpreted as follows: if $[^1A] \to \infty$, so $f_{NR} \to 1$ and $f_R \to 0$; and if $[^1A] \to 0$, so $f_{NR} \to 0$ and $f_R \to 0$, However, Eq. (2.35) shows that the lesser the concentration of ^1A ($[^1A]<<[^1D*]$), the bigger f_R will be, because of $f_R \propto [^1A]$. In other words, to higher concentrations of the acceptor non-radiative energy transfer mechanisms will prevail.

In 1961, Birks and Kuchela evaluated the efficiencies of energy transfer to a host-guest system composed by p-terphenyl (TP) and 1,1,'4,4'-tetraphenyl butadiene (TPB) in toluene solution. Figure 2.6 displays the absorption of an emission spectrum of both molecules and the overlapping spectrum of them. The overlap region, depicted in the dashed line, represents the sum of donor emission and acceptor absorption, which clearly shows the possibility of energy transfer.

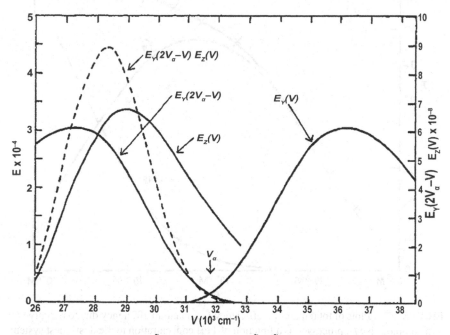

FIGURE 2.6   Absorption $\epsilon_\gamma(\tilde{\upsilon})$ and emission $\epsilon_\gamma(2\tilde{\upsilon}_0 - \tilde{\upsilon})$ spectra of TP and TPB absorption $\epsilon_z(\tilde{\upsilon})$ spectrum (solid lines). TP-TPB overlap spectrum $\epsilon_\gamma(2\tilde{\upsilon}_0 - \tilde{\upsilon}) \epsilon_z(\tilde{\upsilon})$ (dashed line).

Figure 2.7 presents an evolution diagram of energy transfer quantum efficiency, discriminating radiative and non-radiative parts, as well as their sum to the host-guest system TP-TPB as a function of guest concentration (Chen et al., 2011). From the diagram, the overall quantum efficiency was experimentally obtained and adjusted to Eq. (2.34). f_R and f_{NR} were theoretically evaluated.

Initially, for lower concentrations just radiative energy transfer through the trivial mechanism occurs. As the concentration rises and gets bigger than 5 ×

10^{-5} mol L^{-1}, the non-radiative contribution is observed. Then, to concentrations bigger than 5×10^{-3} mol L^{-1}, non-radiative processes are predominant.

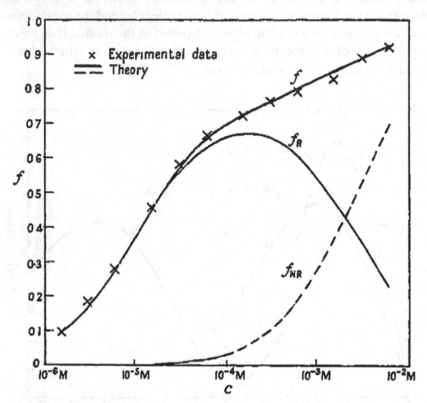

FIGURE 2.7 Plots of total quantum efficiency f_{AD}, quantum efficiency due to radiative f_R, and non-radiative f_{NR} processes as a function of TPB concentration to the host-guest system TP (2.17×10^{-2} mol L^{-1})-TPB.

Another example where the mechanism of radiative energy transfer is beneficial is to the fabrication of OLEDs and PLEDs, which a host-guest system may be employed as a strategy for device construction. In this situation, a photoluminescent material is applied, for instance, as a color-conversion layer (CCL). As previously said, in such cases the CCL must have the absorption spectrum overlapped with the electroluminescence (EL) spectrum of the device active layer. It becomes an elegant way to produce photo/electroluminescent multicolor devices or even white organic light emitting diodes (WOLEDs).

In 2014, Quites et al. described an OLED based on the use of a CCL (Quites et al., 2014a). A polyfluorene derivative (PFOPen) was used as an active layer and a thin film of a poly-p-phenylenevinylene (PPV), the poly[2-methoxy-5-(3′,7′-dimethyloctyloxy)-1,4-phenylene vinylene] (MDMO-PPV) as CCL (acceptor). These two semiconductor polymers present spectral overlap required to trivial energy transfer (Figure 2.8). In this work, the author proposes a combination of the PFOPen EL spectrum, which presents greenish emission, with the MDMO-PPV photolumines-cence (PL) spectrum, which presents an emission on the region of the visible too, but redshifted. The resulting combined-emission was white, with chromaticity coordinates (CIE, 1931) (0.35,0.31)-pure white is on coordinates (0.33,0.33). This system has the advantage of being constituted of a single material, a fact responsible for diminishing other quenching processes, which are a product mainly from morphological defects and compromise the efficiency of the device (Sonawane and Asha, 2011).

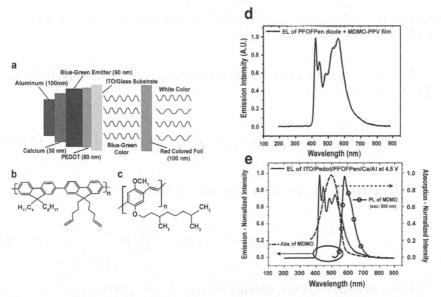

FIGURE 2.8 (a) Schematic view of the radiative energy transfer, (b) PFOPen, and (c) MDMO-PPV structures, (d) normalized EL spectrum of the WOLED, (e) spectral overlap between the PFOPen EL and MDMO-PPV absorption spectra, and the PL spectrum to the CCL. (*Source:* Reprinted from Quites et al., 2014b. © Elsevier.)

2.6.4.2 NON-RADIATIVE ENERGY TRANSFER

In this section, the theoretical and experimental concept regarding a specific mechanism of non-radiative energy transfer related to photophysical and photochemical processes, the Förster resonance energy transfer (FRET) mechanism. A model developed by Theodore Förster during the 1940s (Singh et al., 2016).

The kinetic mechanism to the Förster model is of pseudo-first-order, like collisional quenching (1.2.4.3), as depicted in Figure 2.9. Despite not depending on collisions, FRET is treated as a bimolecular process between $^1D^*$ and ^1A.

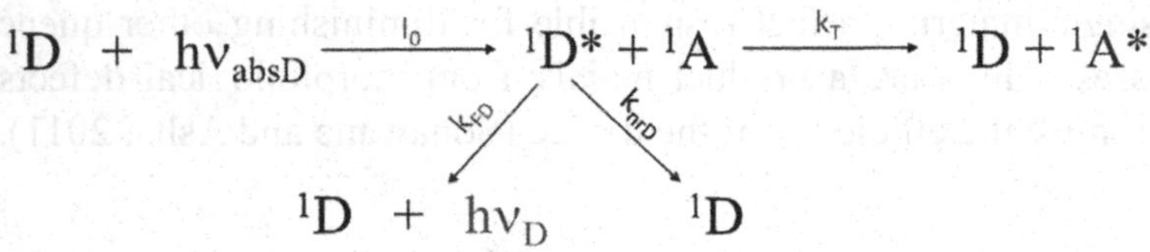

$$^1D \;+\; h\nu_{absD} \xrightarrow{\;I_0\;} {}^1D^* + {}^1A \xrightarrow{\;k_T\;} {}^1D + {}^1A^*$$

$$\underset{k_{FD}}{\swarrow} \qquad \underset{k_{nrD}}{\searrow}$$

$$^1D \;+\; h\nu_D \qquad\qquad {}^1D$$

FIGURE 2.9   FRET kinetics mechanism between donor $^1D^*$ and acceptor ^1A. With k_T as the FRET rate constant.

We can write the kinetic equation that describes the deactivation rate of $^1D^*$ as follows:

$$\frac{d\left[{}^1D^*\right]}{dt} = I_0 - \left(k_{FD} + k_{nrD}\right)\left[{}^1D^*\right] - k_T\left[{}^1A\right]\left[{}^1D^*\right] \qquad (2.36)$$

Considering the pseudo-first order approximation:

$$\frac{d\left[{}^1D^*\right]}{dt} = I_0 - \left(k_{FD} + k_{nrD} + k_T\left[{}^1A\right]\right)\left[{}^1D^*\right] \qquad (2.37)$$

Again, to photodynamic conditions, as $[{}^1D^*] \propto I_D(t)$:

$$\frac{d\left[I_D(t)\right]}{dt} = -\left(k_{FD} + k_{nrD} + k_T\left[{}^1A\right]\right)I_D \qquad (2.38)$$

Integration and subsequent application of the exponential function on Eq. (2.38), results on:

$$I_D(t) = I_{D0}e^{-\left(k_{FD}+k_{nrD}+k_T\left[{}^1A\right]\right)t}$$

(2.39)

It is known that $k_{FD} + k_{nrD} = k_D = \dfrac{1}{\tau_D}$, thus:

$$k_{FD} + k_{nrD} + k_T\left[{}^1A\right] = \frac{1}{\tau_{AD}}$$

(2.40)

When species ${}^1D^*$ be in the presence of acceptor ${}^1A$, the fluorescence lifetime of ${}^1D^*$ (τ_D) will decrease to τ_{AD}, the fluorescence lifetime of ${}^1D^*$ in the presence of ${}^1A$, because of FRET.

In other words, replacing Eq. (2.40) on (2.39) gives:

$$I_{DA}(t) = I_{DA0}e^{-\frac{t}{\tau_{AD}}}$$

(2.41)

Further algebraic modifications on Eq. (2.40) offer the FRET efficiency relation:

$$E_{FRET} = \tau_{AD}k_T\left[{}^1A\right] = 1 - \frac{\tau_{AD}}{\tau_D}$$

(2.42)

Until now, the mechanism (Dexter, 1953) to FRET was described in kinetic terms. However, the real meaning of the FRET rate k_T is not fully explained. To do so, Förster theory to non-radiative energy transfer makes use of quantum and classical mechanics (Singh et al., 2016; Turchetti et al., 2014). Considering a system with ${}^1D^*$ in a certain distance r from ${}^1A$ in a rigid and inert solvent, the electronic energy of the donor will be greater than the one of the acceptors ($E_D > E_A$). Thus, intramolecular non-radiative energy transfer is a resonance phenomenon, which involves transitions on isoenergetic levels of donor and acceptor (Figure 2.10) (Zheng et al., 2013; Rogach et al., 2009). In this point, it is relevant to mention that non-radiative energy transfer to such a system can happen by Coulombic interactions (FRET) or by a redox reaction with electron exchange [Dexter model (Dexter, 1953)] between $^1D^*$ and ^1A.

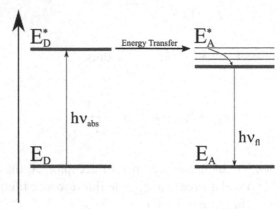

FIGURE 2.10   Schematic view of the FRET process with the energetic levels of donor and acceptor ($E_D | E_\downarrow A^\uparrow$).

Source: From Rogach et al. (2009).

Thereby, the electronic transitions $^1D^* \rightarrow {}^1D \mid {}^1A \rightarrow {}^1A^*$, schematically depicted in Figure 2.9, corresponding to the fluorescence spectrum of $^1D^*$ ($F_D(\lambda)$) and the absorption spectrum of $^1A(F_D(\lambda))$. Furthermore, in terms of the energetic evolution of the process, FRET works as an adiabatic reaction (Turro et al., 1979), and the energy difference $E_D - E_A$ can be partitioned on vibrational levels of the two species (Guo et al., 2013).

Besides, the FRET rate constant is expressed as:

$$k_T = \frac{4\pi p_E}{h} \beta_{el}^2 F \tag{2.43}$$

where, p_E is the density of vibronic state, β_{el} is the matrix element to the electrostatic interaction between initial and final states, h is the Planck constant, and F is the Franck-Condon factor (the overlap integral of vibrational wavefunctions of D and A), given by the sum:

$$F = \sum |\psi_{D^*}| \psi_D \psi_{A^*} | \psi_A |^2 \tag{2.44}$$

where, $\psi_{D*}|\psi_D$, $\psi_{A*}|\psi_A$ are the vibrational wavefunctions of $^1D*|^1D$, $^1A*|^1A$, respectively. Also, the product $\beta^2_{el} F$ is directly related to the spectral overlap $J(\lambda)$ (Guo et al., 2013; Zheng et al., 2013; Föster, 1959).

To be more specific, the Coulombic interaction originated on excited-state between $^1D*$ and ^1A can be divided into two terms: one from the pure electrostatic interaction (defined by Förster, 1959) and another from electron exchange (defined by Dexter in 1951 (Dexter, 1953)). These interactions are the result of the multipole-multipole expansion of charge interaction, with a significant contribution from the dipole-dipole interaction. This "dipole-dipole" force comes from the transaction dipole moments of both species, μ_D and μ_A, and express the term β_{el} by:

$$\beta_{el} \alpha \frac{\mu_D \mu_A}{r^3} \qquad (2.45)$$

With this result, Förster also derived a relation to obtain k_T only expressed by experimental parameters, now as a function of the distance :

$$(2.46)$$

where, n is the medium refractive index, τ_D is the emission lifetime of $^1D*$ in the absence of ^1A; N is the Avogadro constant, k is the orientation factor of the transition dipole moments from $^1D*$ and ^1A, which varies from 0 to 4, the values to perpendicular and parallel transitions, and mathematically expressed by:

$$k = cos\theta_{AD} - 3 \, cos\theta_D \, cos\theta_A \qquad (2.47)$$

where, θ_{AD} is the angle between μ_D and μ_D, θ_D and θ_A are the angles formed between the transaction dipole moments and their resulting vector. k is experimentally obtained by transient anisotropy measurements, but, commonly, it is assumed to be equal to 2/3.

Furthermore, Eq. (2.46) can be rewritten in the most useful way, as follows:

$$k_T(r) = \frac{1}{\tau_D} \left(\frac{R_0}{r} \right)^6 \qquad (2.48)$$

where, R_0 is a critical distance to non-radiative energy transfer, known as the "Förster ratio." It is the distance to an efficiency of energy transfer equal to 50%.

Finally, FRET efficiency E_{FRET} can also be written as a function of the distance between donor and acceptor:

$$E_{FRET}(r) = \frac{R_0^6}{R_0^6 + r^6} \qquad (2.49)$$

FRET mechanism is a strategy widely used in photophysics, photo-chemistry, and photobiology. It became a good strategy for conformational and morphological studies. In 2015, Sardar, and coworkers developed a system based on nanofibers of poly(diphenyl)butadiene (PDPB) and ZnO nanoparticles (NPs) of 5.0 nm in diameter to be applied as an absorptive layer of a hybrid solar cell (Rigby et al., 2012). The use of these two semiconductor materials is interest because of the heterojunction formation, which permits the enhancement of charge separation and extends the spectral absorption region in comparison to the spectrum of the sun.

PDPB was synthesized in a way that several oligomers were produced, resulting in an emission spectrum dependent on the excitation wavelength, once there were many species present. With the ZnO NPs synthesis in the presence of PDPB, a high electronic interaction takes place, and the excitation of PDPB leads to electron transfer to ZnO NPs, which eliminated the excitation wavelength dependency. Moreover, this hybrid organic-inorganic system has a bandgap alignment of both materials also useful in energy transfer through the Förster model. Figure 2.11(b) shows the spectral overlap of ZnO NPs emission and PDPB absorption spectra, along with the decrease of ZnO NPs emission lifetime (Figure 2.11(c) and (d)). P_1 and P_2 refer to the two different species of defective ZnO NPs, Figure 2.11 (a) present the deconvoluted spectrum with their contribution to the overall emission. These two species were assigned by the two lifetimes related to Table 2.1, where one can see that in the presence of PDPB there is a change due to the FRET process, despite experimental errors consequent of the variation of emission wavelength recorded.

From experimental data, the authors could obtain the distance between donor and acceptor as 3.4 nm and 3.1 nm to P_1 and P_2 species, with FRET efficiency 64% and 70%, respectively. It serves as a great example to illustrate potential applications of the FRET process on materials sciences, especially to the case of semiconductor polymers on the construction of optical photovoltaic cells, where FRET becomes a way to energy harvesting.

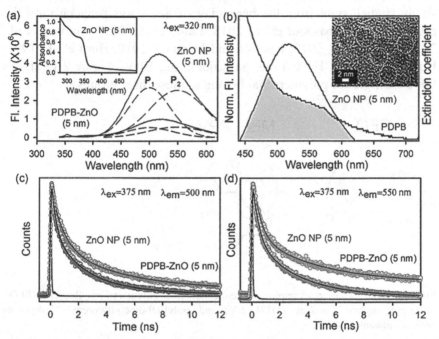

FIGURE 2.11 (a) Emission spectra of ZnO NPs (green) and PDPB-ZnO hybrid material (blue). The inset contains the absorption spectrum of ZnO NPs, (b) the overlap between the emission spectrum of ZnO NPs and the absorption spectrum of PDPB, with the inset containing the high-resolution transmission electron microscopy image of the hybrid material, (c) and (d) show the emission decay of ZnO NPs in the absence (green) and in the presence (blue) of PDPB, recorded at 500 nm and 550 nm, respectively. (*Source:* Reprinted from Sardar et al., 2015. https://creativecommons.org/licenses/by/4.0/)

TABLE 2.1 Emission Lifetimes of ZnO NPs Detected at 500 nm and 550 nm (λ_{em}), with Excitation at 375 nm

Sample	λ_{em} (nm)	τ_1 (ns)	τ_2 (ns)
ZnO NPs	500	0.82	4.22
PDPB-ZnO		0.27	2.15
ZnO NPs	550	0.52	4.01
PDPB-ZnO		0.34	2.63

2.7　EXAMPLES OF SEMICONDUCTIVE POLYMERS PHOTOPHYSICS

Classic examples of SemPolys include the groups of polyfluorenes (PFO), polyparaphenylene (PPV) and polythiophene (PT), which were extensively studied right from the beginning of SemPolys research until the present days (Abbaspoor et al., 2018; Lanzi et al., 2017; Gärtner et al., 2017; Kobe et al., 2013; Faleiros and Miranda, 2010; Burn et al., 1992; Burroughes et al., 1988, 1990; Sauvajol et al., 1989). Figure 2.12 shows three examples of polymers that belong to these three groups.

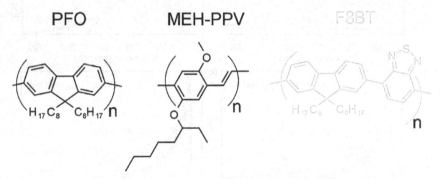

FIGURE 2.12 Classic examples of semiconductive polymers: polyfluorene (PFO), polyparaphenylene vinylene (MEH-PPV) and poly[9,9-dioctylfluorenyl-2,7-diyl]-co-benzothiadiazole (F8BT).

Starting from the blue region of the visible spectrum, PFOs are SemPolys widely used in spectroscopic studies, from optoelectronic devices to bioinspired applications. They are composed of repeated units of fluorene with an alkyl chain at position 9, usually an octyl substituent, resulting on the poly[9,9-dioctylfluorenyl-2,7-diyl] (Figure 2.12, in F8BT structure).

The substituent will be essential to modify polymer solubility, so the polymer backbone is responsible for the emission. Thus, in 2013, Quites, and coworkers developed a study with another PFO derivative, a copolymer based on two fluorene monomers, with an octyl and a pentanyl substituents, named poly[(9,9-dioctylfluorenyl-2,7-diyl)-alt-co-(9,9-di-{5'-pentanyl}-fluorenyl-2,7-diyl)] (PFP) (Quites et al., 2013). Regardless of the alquil chain substitution, PFO have a characteristic emission profile with a vibrionic progression due to the first excited electronic state transition. In the case of PFP, this sequence was found at 419.1 nm, 440.2 nm, and 472.7 nm in THF solution. Figure 2.12 brings the absorption and emission spectra of the polymer, and the inset contains its decay profile, which corroborates with the vibronic progression assignment once it is monoexponential (i.e., there are only one species on the excited state).

The broad and intense absorption band centered at 382 nm are related to π-π* transition, which offered an SS of 2318 cm^{-1} as a consequence of non-radiative deactivations paths. Fluorescence decay produced a lifetime constant of 0.63 ns that along with the quantum yield value (0.33) made possible the application of Eq. (2.22) to discover the natural emission lifetime of 2.03 ns. This vast difference agrees with the SS conclusion and makes sense once one remembers how big a polymer molecule can be, therefore it would be natural to expect several types of inter and intramolecular interactions responsible for quenching the radiative path.

Going to the solid-state, the thin film deposited by casting from THF solution presented not only a broad absorption band, but also a red-edge tail that is the indicative of aggregate formation mainly from π-π stacking. Besides, the emission spectrum continues with the vibronic progression, but redshifted—also a strong indication of intermolecular forces enhancement-and present a small band around 525 nm from aggregated species. These results become valid when analyzed in conjunction with the fluorescence decay: that monoexponential profile seen in solution became biexponential, having as lifetime constants the values 0.53 ns and 0.11 ns. The first one is very similar to the obtained value in the THF solution, so it belongs to the same species of PFP, the second one was attributed to aggregate structures (Figure 2.13).

FRET-like energy transfer was also studied in the SemPolys framework, to understand the exact role of ET processes on organic electronic devices. Nome et al. (2011), described energy transfer from PFO (blue emitter) to MEH-PPV (orange emitter) in solution and solid-state (thin-film) (Figure 2.14), a classic polymer blend to white-light generation, using steady-state, and time-resolved fluorescence measurements. In this manuscript, it was observed that the ET process in this system was composed of two mechanisms: the trivial and FRET. It was possible to separate and quantify these two mechanisms, using the total energy transfer calculation from the steady-state fluorescence spectra and the suppression of the PFO (donor) fluorescence lifetimes in the presence of MEH-PPV (acceptor), quantifying only the FRET mechanism.

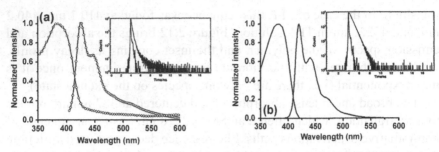

FIGURE 2.13 Absorption (a) and emission (b) electronic spectra of PFP in THF solution. The inset contains the PFP fluorescence decay profile. (*Source:* Reprinted with permission from Quites et al., 2013. © Elsevier.)

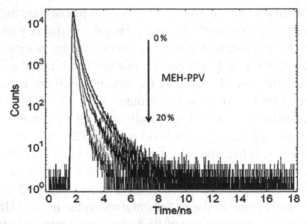

FIGURE 2.14 Fluorescence decays of pure PFO and in blended with MEH-PPV at thin-films (λ_{exc} = 375 nm; λ_{em} = 425 nm). Quencher and FRET-like energy transfer processes from PFO to MEH-PPV. (*Source:* Reprinted with permission from Nome et al., 2011 © Elsevier.)

Trivial and FRET energy transfer processes from PLEDs to dyes CCLs might be used to obtain multicolor and WOLEDs devices, as De Azevedo et al. (2016) and De Azevedo (2017) presented in his two manuscripts. Firstly, the authors introduced the use of PFO and F8BT PLEDs EL emission with Erythrosin B (ERB) and Rose Bengal (RB) environment-friendly composites with poly(vinyl alcohol) (PVA) at (0.01% mol/mol ratios) CCLs (Figure 2.15). ERB and RB electronic absorption spectra did not exhibit spectral overlap with the PFO EL spectrum, while they displayed a higher spectral overlap with F8BT EL spectrum. Thus, combining the red-dyes PL color with the yellow color of the PLED device, tuning the resulting color from yellow to red emission, increasing the PVA: dye film thickness. However, an efficiency loss was observed with the addition of the CCL layers (De Azevedo et al., 2016).

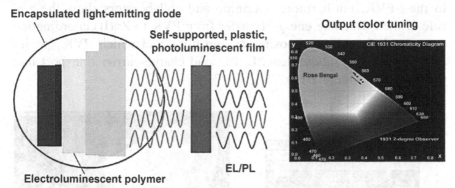

FIGURE 2.15 Trivial energy transfer from F8BT PLEDs to friendly-environment PVA: Reddyes casting films: Color tuning from green to red emission (Source: Reprinted with permission from De Azevedo et al., 2016. © Elsevier).

Now, in their manuscript published in 2017 (Figure 2.16) (De Azevedo et al., 2017), the authors separated the contributions of PFO and F8BT emissions in two kinds of the device: blend and layer PLEDs as a function of its concentration. Also, the authors identified the contribution of FRET-like energy transfer process in these diodes, with the bi-layer device exhibiting the highest FRET efficiency than the blend PLED. This behavior is possible due to the interpenetration between PFO and F8BT layers, leading a high contact among the polymers in comparison to its polymer blend, according to the authors. Thus, steady-state, and transient photophysical properties are fundamental to understand these processes.

The use of the PL techniques is fundamental to comprehend the devices work as well as which polymer phase, organization, and chromophore favor the radiative pathways in the PLED. Thus, Ye et al. (2010) (PL with temperature), Bonon and Atvars (2012) (steady-state measurements with Stern-Volmmer equation) and Germino et al. (2017) (Fluorescence decays) described the three species of PVK on the excited state, leading to a multiexponential fluorescence decay, using steady-state and transient techniques. According to these authors' observations and measurements, PVK polymer presents a multiexponential decay composed of three species: the isolated carbazole (PVK), the partially overlapped (p-PVK), and the fully overlapped (f-PVK) species. According to the author's observations, the fastest fluorescence lifetime was attributed to isolated carbazole, the intermediary to the p-PVK, and the slowest to the f-PVK. Furthermore, Germino and collaborators show the real role of the FRET-like energy transfer from PVK to Zn(II) coordination compounds into its composites at thin-films and which PVK species contribute Zn(II) salicylidenePL, EL, and charge-carrier transport into their devices (Figure 2.17).

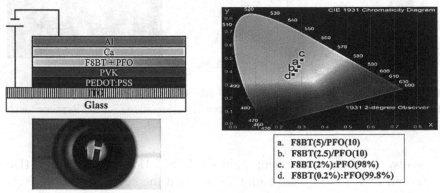

a.	F8BT(5)/PFO(10)
b.	F8BT(2.5)/PFO(10)
c.	F8BT(2%):PFO(98%)
d.	F8BT(0.2%):PFO(99.8%)

FIGURE 2.16 Correlation between the PL and EL emissions of polyfluorene-based PLEDs using bilayers and polymer blends and the impact of the ET processes to color tuning and device efficiencies (Source: Reprinted with permission from De Azevedo et al., 2017. © Elsevier).

2.8 CONCLUDING REMARKS

Finally, this chapter presented the main concepts of photophysical processes that could be observed in organic chromophores, including conducting and semiconducting polymer systems. Exciton model was briefly introduced to understand the formation of electronically excited states in luminescent polymers and their return to electronic ground states due to radiative and non-radiative processes. Quenching processes by collisional model, inner filter, trivial, and resonant energies transfer was explained so the readers feel able to understand the essential concepts that might influence on conducting polymer devices, such as photovoltaic and OLEDs.

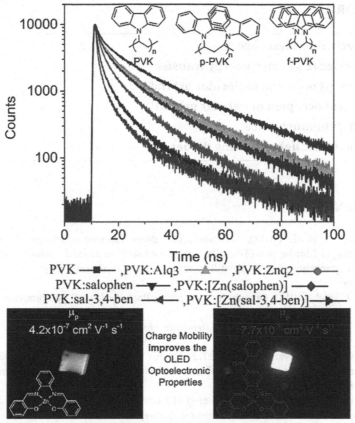

FIGURE 2.17 Fluorescence decays of PVK and its coordination compounds thin-films composites, presenting resonant energy transfer from PVK to the dyes and which PVK species contribute to the FRET-like process. Also, its correlation to the charge transport properties of their OLEDs. (*Source:* Adapted from Germino et al., 2017.)

Also, current examples of the photophysical importance and application to solve some problems of the organic light-emitting diodes (OLED) properties, for example, energy transfer processes, dynamic, and static quenching, aggregation, excimers, and exciplex formation, were presented.

Fluorescence sensitivity is so high that it provides the needed tools to explain how molecular systems behave on an electronically excited state and allows the construction of environmentally friendly devices, minimizing the use of metals and energy saving.

KEYWORDS

- electroluminescence
- Förster resonance energy transfer
- highest occupied molecular orbital
- lowest occupied molecular orbital
- photoluminescence
- polymeric light-emitting diodes

REFERENCES

Abbaspoor, S., et al., (2018). Supramolecular donor-acceptor structures via orienting predeveloped fibrillar poly(3-hexylthiophene) crystals on bared/functionalized/grafted reduced graphene oxide with differentthiophenic constituents. *Org. Electron.*, *52*, 243–256.

Baeg, K. J., et al., (2013). Organic light detectors: Photodiodes and phototransistors. *Adv. Mater.*, *25*(31), 4267.

Birks, J. B., (1970). *Photophysics of Aromatic Molecules* (1ˢᵗ ed.). John Wiley & Sons Ltda.

Bonon, B. M. A., &Atvars, T. D. Z., (2012). Energy transfer from poly(vinyl carbazole) to a fluorene-vinylene copolymer in solution and in the solid state. *Photochem. Photobiol.*, *88*(4), 801–809.

Burn, P. L., et al., (1992). Chemical tuning of electroluminescent copolymers to improve emission efficiencies and allow patterning. *Nature*, *356*(6364), 47.

Burroughes, J. H., et al., (1988). New semiconductor device physics in polymer diodes and transistors. *Nature*, *335*(6186), 137.

Burroughes, J. H., et al., (1990). Light-emitting diodes based on conjugated polymers. *Nature*, *347*(6293), 539.

Chen, W. H., et al., (2011). A highly selective pyrophosphate sensor based on ESIPT turn-on in water. *Org. Lett.*, *13*(6), 1362.

De Azevedo, D., et al., (2016). Tuning the emission color of a single-layer polymer light-emitting diode with a solution-processed external layer. *Synth. Met.*, *222*, 205.

De Azevedo, D., et al., (2017). Correlation between the PL and EL emissions of polyfluorene-based diodes using bilayers or polymer blends. *Synth. Met.*, *233*, 28.

Dexter, D. L., (1953). A theory of sensitized luminescence in solids. *J. Chem. Phys.*, *21*(5), 836.

Facchetti, A., (2011). *Chem Mater.*, *23*(3),733.

Faleiros, M. M., & Miranda, P. B., (2010). Photoluminescence of MEH-PPV with ultraviolet excitation. *Synth. Met.*, *160*(23/24), 2409–24012.

Farinola, G. M., &Ragni, R., (2011). Electroluminescent materials for white organic light emitting diodes. *Chem. Soc. Rev.*, *40*(7), 3467.

Förster, T., (1959). 10th spiers memorial lecture. Transfer mechanisms of electronic excitation. *Discuss. Faraday Soc., 27*(10), 7.

Gärtner, S., et al., (2017). Relating structure to efficiency in surfactant-free polymer/fullerene nanoparticle-based organic solar cells. *ACS Appl. Mater. Interfaces, 9*(49), 42986.

Germino, J. C., et al., (2017). Revealing the dynamic of excited state proton transfer of a π-conjugated salicylidene compound: An experimental and theoretical study. *J. Phys. Chem. C., 121*(2), 1283–1290.

Guo, X., et al., (2013). Designing π-conjugated polymers for organic electronics. *Prog. Polym. Sci., 38*(12), 1832.

Ito, A., et al., (2010). A ratiometric TICT-type dual fluorescent sensor for an amino acid. *Phys. Chem. Chem. Phys., 12*, 6641.

Kang, H., et al., (2016). Bulk-heterojunction organic solar cells: Five core technologies for their commercialization. *Adv. Mater., 28*(36), 7821.

Kobe, H., et al., (2013). Photoluminescence characterization of polythiophene films incorporated with highly functional molecules such as metallophthalocyanine. *J. Vac. Sci. Technol. A., 31*(1), 011504.

Lakowicz, J. R., (2006). In: Lakowicz, J. R., (ed.), *Principles of Fluorescence Spectroscopy.* Springer US: Boston, MA.

Lanzi, M., et al., (2017). Water-soluble polythiophenes as efficient charge-transport layers for the improvement of photovoltaic performance in bulk heterojunction polymeric solar cells. *Eur. Polym. J., 97*, 378–388.

Laquai, F., et al., (2009). Excitation energy transfer in organic materials: From fundamentals to optoelectronic devices. *Macromol. Rapid Commun., 30*(14), 1203.

Lerch, M. M., et al., (2016). Orthogonal photoswitching in a multifunctional molecular system. *Nat. Commun.,* 7, 12054.

Luhman, W. A., & Holmes, R. J., (2011). Investigation of energy transfer in organic photovoltaic cells and impact on exciton diffusion length measurements. *Adv. Funct. Mater., 21*(4), 764.

Mazzio, K. A., & Luscombe, C. K., (2015). The future of organic photovoltaics. *Chem. Soc. Rev., 44*(1), 78.

McQuarrie, D. A., & Simon, J. D., (1997). *Physical Chemistry: A Molecular Approach* (p. 1360). University Science Books.

Nome, R. A., et al., (2011). Electronic energy transfer between poly(9,9'-dihexylfluorene-2,2-dyil) and MEH-PPV: A photophysical study in solutions and in the solid state. *Synth. Met., 161*(19–20), 2154.

Orgiu, E., &Samorì, P., (2014). 25th anniversary article: Organic electronics marries photochromism: Generation of multifunctional interfaces, materials, and devices. *Adv. Mater., 26*(12), 1827.

Quites, F. J., et al., (2013). Facile control of system-bath interactions and the formation of crystalline phases of poly[(9,9-dioctylfluorenyl-2,7-diyl)-alt-co-(9,9-di-{5'-pentanyl}-fluorenyl-2,7-diyl)] in silicone-based polymer hosts. *Eur. Polym. J., 49*(3), 693.

Quites, F. J., et al., (2014a). Improvement in the emission properties of a luminescent anionic dye intercalated between the lamellae of zinc hydroxide-layered. *Colloids Surf. A Physicochem. Eng. Asp., 459*, 194.

Quites, F. J., et al., (2014b). White emission in polymer light-emitting diodes: Color composition by single-layer electroluminescence and external photoluminescence component. *Mater. Lett., 130*, 65.

Reineke, S., et al., (2013). White organic light-emitting diodes: Status and perspective. *Rev. Mod. Phys., 85*(3), 1245.

Reiss, P., et al., (2011). Conjugated polymers/semiconductor nanocrystals hybrid materials-preparation, electrical transport properties, and applications. *Nanoscale, 3*(2), 446.

Rigby, N., et al., (2012). Metal complexes of 2,2′-Bipyridine-4,4′-diamine as metallo-tectons for hydrogen bonded networks. *Cryst. Growth Des., 12*(4), 1871.

Rogach, A. L., et al., (2009). Energy transfer with semiconductor nanocrystals. *J. Mater. Chem., 19*(9), 1208.

Rohatgi-Mukherjee, K. K., (1978). *Fundamentals of Photochemistry* (p. 347). New Age International.

Sardar, S., Kar, P., Remita, H. et al. Enhanced Charge Separation and FRET at Heterojunctions between Semiconductor Nanoparticles and Conducting Polymer Nanofibers for Efficient Solar Light Harvesting. Sci Rep 5, 17313 (2015). https://doi.org/10.1038/srep17313

Sauvajol, J. L., et al., (1989). Photoluminescence in polythiophene and polyselenophene. *Synth. Met., 28*, 293.

Singh, A., et al., (2016). Selective detection of Hg(II) with benzothiazole-based fluorescent organic cation and the resultant complex as a ratiometric sensor for bromide in water. *Tetrahedron, 72*(24), 3535.

Sonawane, S. L., & Asha, S. K., (2016). Fluorescent polystyrene microbeads as invisible security ink and optical vapor sensor for 4-nitrotoluene. *ACS Appl. Mater. Interfaces, 8*(16), 10590.

Strickler, S. J., & Berg, R. A., (1962). Relationship between absorption intensity and fluorescence lifetime of molecules. *J. Chem. Phys., 37*(4), 814.

Turchetti, D. A., et al., (2014). A photophysical interpretation of the thermochromism of a polyfluorene derivative-europium complex. *J. Phys. Chem. C., 118*(51), 30079.

Turro, N. J., et al., (1979). Adiabatic photoreactions of organic molecules. *Angew. Chemie Int. Ed., 18*(8), 572.

Uoyama, H., et al., (2012). Highly efficient organic light-emitting diodes from delayed fluorescence. *Nature, 492*(7428), 234.

Valeur, B., (2002). *Molecular Fluorescence: Principles and Applications* (pp 20–33). Wiley-VCH Verlag GmbH, Weinheim.

Van Roosbroeck, W., & Shockley, W., (1954). Photon-radiative recombination of electrons and holes in germanium. *Phys. Rev., 94*(6), 1558.

Vezie, M. S., et al., (2016). Exploring the origin of high optical absorption in conjugated polymers. *Nat. Mater., 15*(7), 746.

Ye, T., et al., (2010). Electroluminescence of poly(N-vinylcarbazole) films: Fluorescence, phosphorescence, and electromers. *Phys. Chem. Chem. Phys., 12*(47), 15410.

Zheng, H., et al., (2013). Genetic design of enhanced valley splitting towards a spin qubit in silicon. *Nat. Commun., 4*, 1.

CHAPTER 3

Theoretical Aspects of Semiconducting Polymers

CRISTINA A. BARBOZA[1] and RODRIGO A. MENDES[2]

[1]Institute of Physics, Polish Academy of Sciences, 02 668, Warsaw, Poland

[2]Biophotonics Laboratory, CePOF-IFSC/USP, São Carlos, Brazil

3.1 INTRODUCTION

The Schrödinger equation gives the physical laws governing the behavior of the material, and the common goal of conventional *ab initio* methods consists in solving it within Born-Oppenheimer approximation:

$$H\Psi = E\Psi \tag{3.1}$$

However, it is not possible to explain it directly for realistic systems, since Ψ is an N-particle wavefunction such as semiconducting polymers (SemPolys) (Schrödinger, 1926). Over decades, several methods were developed to try to find the best ratio between accuracy and computational cost required for the quantum mechanical calculations, among them can be cited: (1) Semiempirical (Stewart, 1989a,b, 1991, 2007, 2013) which use parameters derived from experimental data to simplify calculations. (2) Hybrid procedures such as quantum mechanics/molecular mechanics (QM/MM) (Senn and Thiel, 2009; Svensson et al., 1996) to compute properties of large molecular systems, where a crucial part of the system is treated explicitly by a quantum mechanical level whereas the rest of the system is approximated by a classical or MM force field treatment. (3) For reduced models such as monomers, post-Hartree-Fock (HF) approaches such as Møller-Plesset (MP2) (Cremer, 1998), complete active space second-order perturbation theory (CASPT2) (Burke, 2012) and coupled-cluster singles and doubles (CCSD) (Gaydaenko and Nikulin, 1970) have also been used. These methods offer a significant improvement in the accuracy; however, they scale-up to a fifth or even higher power with the size of the system.4. Density functional theory (DFT), which in the last few decades DFT emerged as a practical theory, available for 'everyday use' and will be discussed in more detail in the following sections (Jain et al., 2016; Salzner, 2014).

3.2 HISTORICAL BACKGROUND

The development of the DFT has following an exciting path, being the subject of multiple debates and controversies among researchers. However, today is an indispensable tool for most applications in the field of theoretical physics and chemistry (Kryachko and Ludeña, 2014). Hohenberg and Kohn developed this method, firstly reported in their article entitled *"Inhomogeneous Electron Gas"* (Hohenberg and Kohn, 1964), yielding the so-called Hohenberg-Kohn (HK) theorem. At the year of 1965, Kohn and Sham published the seminal article to DFT, called *Self-Consistent Equations Including Exchange and Correlation Effects*. They proposed three main equations that became known as Kohn-Sham (KS) equations, which could be solved iteratively through the self-consistent field method. DFT reformulates the Schrödinger equation, which describes the behavior of electrons in a system, such that approximate solutions are tractable for realistic materials. For this achievement, Walter Kohn received the Nobel Prize in Chemistry in 1998 (Kohn, 1999).

Although DFT is in principle exact, in practice approximations must be made for how electrons interact with each other. These interactions are approximated with so-called exchange-correlation (XC) functionals. Shortly after that DFT discovery, the simplest XC called local density approximation (LDA) was proposed, and the first calculations using it were done by Tong and Sham in 1966. In 1967, articles reporting calculations on solids using the Gáspár-Kohn-Sham potential for exchange (with and without a correlation potential) appeared, and the articles of Kohn and Sham started to be frequently cited (Corminboeuf et al., 2006). LDA became the popular standard in calculations in solids in the 1970s and

1980s (Burke, 2012). However, it is only in the 1970s that local correlation functionals for calculations on solids in the framework of KS equations started to be developed. In the field of quantum chemistry, the first application of the density functional concept was made at the beginning of a decade of 1970 by Gaydaenko and Nikulin (1970), and Gordon and Kim (1972), using the Xα approximation. These authors considered rare gas and ion-ion interactions, considering three-body effects, including gradient corrections to XC functional, which led to several difficulties. Hence, in the middle of the 1970s, both the successes and limitations of this method started to be noticed.

In 1976, Gunnarsson and Lundqvist showed that the LDA method was able to describe molecular bonding in diatomic systems such as H_2 more accurately than the Thomas-Fermi theory. The 1980s have witnessed continuous developments of DFT, since molecules in LDA are typically overbound by about 1eV/bond, and in the late 1980s, generalized gradient approximations (GGAs) produced an accuracy that was useful in chemical calculations (Perdew et al., 1992). The so-called generalized approximation of Wang and Perdew (1991) initiated then a long series of gradient corrected functionals, which led to much more certain energetic properties, and the PBE (Perdew et al., 1996) GGA has come to dominate applications to periodic systems (materials). In the early 1990s, density functionals referred to as hybrids were introduced by Becke, replacing a fraction of GGA exchange with HF exchange, leading to the ubiquitous B3LYP (Becke, 1993), which together to LDA and PBE dominate the user market today (Peverati and Truhlar, 2014).

Much of the success of the DFT comes from the fact that XC functionals often yield accurate results for organic and inorganic systems. A significant contribution was performed in 1985 by Car and Parrinello in the form of an elegant molecular dynamics scheme based on DFT. On the early 1990s, the first DFT codes became broadly available, such as Gaussian (Frisch et al., 2009) software and Amsterdam Density Functional (ADF) (Te Velde et al., 2001) code, what contributed considerably to the extensive use of DFT in modeling and simulation applications. Numerous articles and textbooks are reviewing the extensive use of DFT on the 1990s and 2000s (Kryachko, 2014; Cohen et al., 2012; Capelle, 2006). Nowadays DFT is the most popular approximation to study properties of materials, mainly due to its balance of accuracy and applicability (Geerlings et al., 2003; Baerends and Gritsenko, 1997).

The choice of XC functionals determines, to a large extent, the accuracy of the DFT calculation. Although theorists can often improve the efficiency of the prediction (typically at higher computational cost) using more complex functionals, highly correlated electron systems exist for which most functionals fail. Other limitations of standard DFT include the small system size, difficulty in modeling weak interactions, such as van der Waals, dynamics over long periods, and excited-state properties. However, methods for overcoming these limitations are often available, for instance, larger systems can be treated using the linear scaling approximation, and finite-temperature effects can be addressed through lattice dynamics and cluster expansion. Considering that most of the standard density functionals are unable to describe long-range electron correlations properly, and adequately describe weak interactions such as Van der Waals, hydrogen bonds and π-stacking (Grimme et al., 2010; Gräfenstein and Cremer, 2009), present in systems such as biomolecules and molecular crystals, recently Grimme and coworkers (Grimme, 2006) proposed a scheme called density functional dispersion correction, which mixes conventional functionals and an add-on energy term. The inclusion of the dispersion correction energy term is computationally inexpensive and can lead to significant improvements in the accuracy of the calculations.

Besides, on the numerous applications of the ground state DFT, in 1984, Erich Runge and E.K.U Gross proposed, in the article called *Density-Functional Theory for Time-Dependent Systems,* a time-dependent extension of Hohenberg Kohn theorem. It states that applying a perturbation (Runge and Gross, 1984) from an external potential in a given state, it is possible to produce a time-dependent wavefunction, as a consequence in a second stage it is possible to generate the time-dependent density and is known as the Runge-Gross theorem. A few years later, the Van Leeuwen theorem arise in 1999, expanding the initial TD-DFT proposal to the interaction of too many body systems. TD-DFT is most used to spectroscopic scenarios, i.e., excited states properties, absorption, emission, partial atomic charges, dipole moments and electron densities, and have been extensively used in the field of organic electronics, materials science, among others in the last past decades (Laurent and Jacquemin, 2013; Jacquemin et al., 2011).

DFT and its time-dependent framework are powerful tools enabling theoretical prediction of the molecular geometries, energy levels, and absorption spectra of conjugated organic molecules such as homopolymers, donor-acceptor (D-A) polymers, with reasonable accuracy at affordable computational costs. However, performing DFT calculations on conjugated polymers is still challenging due to many atoms and lack of symmetry. To a first approximation, properties of anisotropic COPs can be investigated on isolated polymers or oligomers. For separate species, DFT and its time-dependent counterpart (TD-DFT) have led to valuable insights and accurate predictions. The structure of the polymers can be approximated using an oligomer approach, and intermolecular interactions can be assessed with three-dimensional band structure calculations or via cluster models. Ultimately, predicting device performance requires also modeling of morphology, such as crystal structures including their defects as well as charge carrier dynamics at interfaces. These are fields of research that are gaining momentum nowadays.

3.3 GENERAL FRAMEWORK

3.3.1 DENSITY FUNCTIONAL THEORY (DFT)

Ground state DFT is based on the HK theorems, which states that there is a one-to-one correspondence between the ground state density and the external potential and that the exact ground state density of a system in an external potential can be found by minimizing an energy functional (Kryachko and Ludeña, 2014). The precise form of this XC function remains unknown. However, approximations to it based on electron gas models and further extensions have proved successful for many classes of materials, facilitated by the exponential increase in computing power in the last few decades.

Walter Kohn and Pierre Hohenberg (Kohn, 1999; Hohenberg and Kohn, 1964), based on the approximation of Thomas-Fermi, proposed that all the information of any system in the ground state could be obtained from the electronic density of these systems. The normalized density determines the total number of electrons in the system, as shown in the following equation:

$$N = \int \rho(r) dr$$

(3.2)

In its first part, the HK theorem defines a universal functional f $[\rho(r)]$, which may take the form of any observable (O) of a system in its non-degenerate ground state, regardless of its external potential. Their HK theorem advances naturally to its second part where the energy functional appears, which can (respecting the variational theorem) represent the ground state energy of the studied system. This theorem is based on the idea that given an electronic density of a system in its ground state $\rho_0(r)$, it is possible to calculate the correspondent ground state wave function $\psi_0(r_1, r_2..., r_N)$, and vice-versa. Hence, these two functions contain the same amount of information. Mathematically, this finding suggests that ψ_0 can be represented as a functional of ρ_0, as follows below:

$$\psi_0\left(r_1, r_2 ..., r_N\right) = \psi_0\left[\rho_0\left(r\right)\right] \qquad (3.3)$$

which means that all observables of the system ground state could be written as density functional as well, as shown in the following equation:

$$O_0 = O\left[\rho_0\right] = \psi\left[\rho_0\right]\hat{O}\,\psi\left[\rho_0\right] \qquad (3.4)$$

It must be considered that the ground state wave function needs to reproduce the electronic density and minimize the energy, hence;

$$E_{v,0} = \min{}_{\psi \to \rho_0}\left\langle \psi \left| \hat{T} + \hat{U} + \hat{V} \right| \psi \right\rangle \qquad (3.5)$$

where, $E_{v,0}$ stands for the ground state energy at a potential $v(r)$, $\hat{T}$ is the kinetic energy operator, the operator $\hat{U}$ represents the electron-electron interaction potential, $\hat{V}$ is the operator that describes the electron-nucleus interaction. Hence, for an arbitrary density, we define the functional:

$$E_v[\rho] = \min{}_{\psi \to \rho_0}\left\langle \psi \left| \hat{T} + \hat{U} + \hat{V} \right| \psi \right\rangle \qquad (3.6)$$

If ρ is a different density from that of the ground state (ρ_0) at the potential $v(r)$, then the wave function that produced this density is different from ψ_0 as well. According to the variational principle, the minimum obtained from $E_v[\rho]$ must be equal or higher than $E_{v,0} = E_v[\rho_0]$. Thus, the functional $E_v[\rho]$ will be minimized by ρ_0 and its value at the minimum point will be $E_{v,0}$. The total energy functional can be written as:

$$E_v[\rho] = \min_{\psi \to \rho_0} \left\langle \psi \left| \hat{T} + \hat{U} + \right| \psi \right\rangle + \int \rho(r) v(r) d^3 r = F[\rho] + V[\rho] \quad (3.7)$$

where, $F[\rho]$ is a universal functional and has the form of $F[\rho] = \min_{\psi \to \rho} \left\langle \psi \left| \hat{T} + \hat{U} \right| \psi \right\rangle$ and is independent of N and $v(r)$. This functional can be rewritten as:

$$F_{HK}[\rho] = T[\rho] + U[\rho] \quad (3.8)$$

where, the subscript HK is due to this function being known as HK functional. Still, Eq. (3.7) suggests that external potential can be written as:

$$V[\rho] = \int \rho(r) \, v(r) \, d^3 r \quad (3.9)$$

Depending on the $v(r)$, which means depends on the system under investigation, making it non-universal.

Because of what has been demonstrated above, HK theorem can be usually partitioned into two theorems, which can be exposed as follows: Theorem I: *"The external potential v(r) and consequently, the total energy is a unique functional of the electron density.* Theorem II: *The total ground state energy (E[ρ_0]) can be reached variationally as a density functional, for this, the density responsible for minimizing E[ρ_0] is the ground state density."*

Hohenberg and Kohn (1964) proved that critical information could be obtained through electronic density; however, they did not demonstrate in practice how this could be done. Walter Kohn and Lu Jeu Sham did this about one year later (Bickelhaupt et al., 2007; Kohn, 1999; Kohn and Sham, 1965), thus solidifying the DFT.

If the external potential of Eq. (3.6) is fixed, it is possible to rewrite the energy functional as shown below:

$$E[\rho] = F_{HK}[\rho] + \int \rho(r) \, v(r) \, d^3 r \quad (3.10)$$

Now, the energy functional is independent of $v(r)$, resulting in a universal form of the density functional. The universal functional is, however, not known and practical DFT involves finding approximate forms for the functional. In the KS, approach the functional splits as:

$$F_{HK}[\rho] = T_S[\rho] + \frac{1}{2}\int\int drdr' \frac{\rho(r)\rho(r')}{|r-r'|} + E_{xc}[\rho] \tag{3.11}$$

where, the first term $T_S[\rho]$ is the kinetic energy of a non-interacting system, the second term is the Coulombic interaction and the last term is the unknown part called the XC functional. The central assumption is that for any ground-state density of an interacting system there exists a non-interacting system with the same ground-state density. The ground state energy can be found by minimizing the energy functional as follow:

$$0 = \frac{\partial}{\partial\rho(r)}\{E[\rho] - \mu\int\rho(r)dr\} = \frac{\partial T_S\rho(r)}{\partial\rho(r)} + v(r) + \int\frac{\rho(r')}{r-r'}dr' + v_{xc}[\rho](r) - \mu \tag{3.12}$$

With the Lagrange multiplier μ ensuring the correct number of electrons. The functional derivative of the unknown XC-functional has been introduced as:

$$v_{xc}[\rho](r) = \frac{\partial E_{xc}[\rho]}{\partial\rho(r)} \tag{3.13}$$

Hence, for a system of non-interacting particles moving in an effective potential v_{eff} we have:

$$v_{eff} = v(r) + \int\frac{\rho(r')}{|r-r'|}dr' + v_{xc}[\rho](r) \tag{3.14}$$

Therefore, it is possible to find the ground state density of the interacting system by solving a set of effective one-particle equations using the KS equations given by:

$$\left[-\frac{\hbar^2\nabla^2}{2m} + v_{eff}(r)\right]\phi_i(r) = \varepsilon_i\phi_i(r) \tag{3.15}$$

where, $\phi_i(r)$ is a KS orbital and ε_i is the corresponding energy. The density of the system is given as the sum of the occupied KS orbitals by:

$$\rho(r) = \sum_{i=1}^{N} f_i |\phi_i(r)|^2 \qquad (3.16)$$

with f_i as the occupation number. If the XC-potential is exact, the density will be the exact density of the interacting system. Therefore, in the KS approach, the problem of finding an approximation to the universal functional $F_{HK}[\rho]$ is reduced to finding an approximation to the XC-functional.

3.3.2 TIME-DEPENDENT DENSITY FUNCTIONAL THEORY (DFT)

The time-dependent extension of the KS equations was derived by Runge and Gross theorem. As the example of the ground state DFT, it was assumed that a potential $v_s(r,t)$ exists and it can reproduce the time-dependent density of the interacting system. TD-DFT describes the linear (first-order) response of the electron density of a non-interacting KS system onto an external perturbation by an oscillating electric field, which, after Fourier transformation, yields excitation energies and transition dipole moments. The density of the interacting system can be found through:

$$i\frac{\partial}{\partial t}\phi_i(r,t) = \left(-\frac{\nabla^2}{2} + v_{eff}[\rho](r,t)\right)\phi_i(r,t) \qquad (3.17)$$

where, the time-dependent KS potential is written as:

$$v_{eff}[\rho](r,t) = v(r) + v^{per}(r,t) + \int \frac{\rho(r')}{|r-r'|}dr' + v_{xc}(r,t) \qquad (3.18)$$

where, $v(r)$ is the field of the nuclei, $v^{per}(r,t)$ is an external time-dependent perturbation, the third term is the Coulomb term and the last one is the time-dependent XC potential. The time-dependent XC potential, $v_{xc}(r,t)$ is usually adopted within the adiabatic approximation which considers $v_{xc}(r, t) \approx v_{xc}(r)$. The time-dependent density is then given by the sum of the occupied time-dependent KS orbitals such as:

$$\rho(r,t) = \sum_{i=1}^{N} f_i |\phi_i(r,t)|^2 \qquad (3.19)$$

It extends the concept of stationary DFT to time-dependent situations: For any interacting quantum, many-particle systems subject to a given time-dependent potential all physical observables are uniquely determined by knowledge of the time-dependent density and the state of the system at an arbitrary, single instant in time. In this case, the initial state of the system at time t_0 is a unique function of the density at this moment which is identical with the ground state density of the stationary system that one has before t. This unique relationship allows deriving a calculation scheme in which a density-dependent single-particle potential represents the effect of the particle-particle interaction.

3.4   ELECTRONIC STRUCTURE OF CONDUCTING ORGANIC POLYMERS

Conducting polymers are a class of materials exhibiting semiconducting and, in some cases, metallic behavior. Delocalization of the π-electrons in the framework of these polymers gives rise to semiconductor-like energy bands and electric conductivity; once mobile charge carriers are produced by electronic excitation, oxidation, or reduction. The resulting electrical conductivity depends on several factors such as their oxidation level, chain alignment, interchain reactions, conjugation length, and degree of disorder, among others (Bubnova et al., 2014).

There is a large variety of semiconductor polymers, and modifications in molecular structure influence their conducting properties and the performance of electronic devices, such as photovoltaic cells and electrochemical supercapacitors (Heeger, 2010). For these materials, the magnitude of the highest occupied molecular orbital (HOMO), lowest unoccupied molecular orbital (LUMO) energy gap is a crucial parameter when determining possible applications of conducting polymers, since it directly affects the short circuit current, once the HOMO energy of the electron donor moiety relative of the HOMO of the electron acceptor unit is proportional to the open-circuit voltage, and the offset between the LUMO of the donor and acceptor defines if the charge separation at the donor/acceptor interface would occur, leading to a photocurrent (Wykes et al., 2013).

Although HOMO energies from B3LYP have been accepted as good predictors for oxidation potentials, the calculated LUMO energy of homopolymers and D-A polymers is consistently overestimated, in some cases by ~1.0 eV (Pastore et al., 2010). Besides, other errors are observed, such as the polymer bond length alternation, ionization potentials, electronic transitions as a function of size, and overestimation of molecular orbitals delocalization (Oliveira et al., 2016). Considering a balance between computational cost and accuracy, a better description of these properties can be obtained by using density functionals containing ~30% of HF exchange, and the use of a long-range corrected such as CAM-B3LYP can improve these results (Wykes et al., 2013) significantly. However, functionals with higher HF exchange, such as M06HF (100% HF exchange), have shown good results in the prediction of vertical, adiabatic excitations or charge carriers for -conjugated systems (Oliveira et al., 2016; Wykes et al., 2013).

Besides, on the usual DFT approach, to evaluate the electronic properties of materials such as organic molecules and small clusters TD-DFT guarantees reliable predictions (Salzner, 2014). Excitation energies can be obtained using two approaches: (i) vertical energies, disregarding structural changes due to the photon absorption process or (ii) adiabatic energies obtained from the difference of energy between optimized geometries of the ground and excited states. Due to the impossibility to treat the whole polymer structure at this level of calculation, a reduced number of oligomers can be used for the calculation of the interest properties and extrapolated to ideal infinite polymers, through the so-called oligomer approximation. In this case, to choose a proper theoretical extrapolation method is mandatory (Larsen, 2016; Müllen and Wegner, 2008). Reliable results have been obtained using TD-DFT to describe the lowest optically allowed the excited state of conjugated chains (Gierschner et al., 2007).

From the theoretical point of view besides their light absorption and emission profiles, the most exciting properties of these materials are ionization potential (I) and electron affinity (Ea). The relationship between electron binding energies and orbital energies of the neutral species as obtained with band structure calculations is provided by the Koopman's theorem (Manne and Aberg, 1970), which states that negative HF orbital energies are the ionization potential (I) and electron affinities (A) of a molecule. However, within the KS formalism, the meaning of eigenvalues is given by the Janak theorem:

$$\frac{\partial E_{v_{ext}}}{\partial n_j} = \epsilon_j \qquad (3.20)$$

where, n_j is the occupation number of the j_{th} orbital. Hence, for the exact E_{xc} functional, HOMO, and LUMO can be written as:

$$\epsilon_N^{KS}(N) = -I \qquad (3.21)$$

and

$$\epsilon_{N+1}^{KS}(N+1) = -A \qquad (3.22)$$

Hence, the Janak theorem shows the connection between the quantities I and A and the energies of the HOMO and LUMO, respectively. Higher ionization energies can also be obtained using TD-DFT, however, the multiconfigurational character of the involved transition increases with the energy of the transition, and DFT/TD-DFT are monodeterminantal methods of calculation, and hence the error in the predicted energies would be much higher than the ones respective to the first ionization potential, even using large basis sets.

Ionization potentials and electron affinities of π-conjugated organic oligomers are well described within the DFT framework. For instance, Salzner and Aydin (2011) suggested that the use of γ-tuned range-separated functionals, which allows the control at which distance from the nucleus the amount of HF exchange starts to increase, would solve the incorrect chain length dependence of first ionization potentials, excitation energies, and orbital energies, and among these functionals, those with better performance are wB97X, wB97XD, and CAMB3LYP.

3.5   CONCLUDING REMARKS AND PERSPECTIVES

In this chapter, the development, theoretical foundation, and the performance of the DFT and its time-dependent counterpart to describe the electronic properties of conducting polymers were briefly discussed. Due to its excellent balance between accuracy and computational cost for the calculations of oligomers, DFT, and the density functional B3LYP, became the most popular approximation to study of electronic properties like ionization potentials, electron affinities and excitation energies of these materials, contributing to the design of new conducting polymers for electronic devices. However, new density functionals and extensions to the DFT are still being developed, such as the DFT-D scheme and the linear scaling TD-DFT, among others.

KEYWORDS

- **complete active space second-order perturbation theory**
- **coupled-cluster singles and doubles**
- **density functional theory**
- **local density approximation**
- **Møller-Plesset**
- **quantum mechanics/molecular mechanics**

REFERENCES

Baerends, E. J., & Gritsenko, O. V., (1997). A quantum chemical view of density functional theory. *The Journal of Physical Chemistry A, 101*(30), 5383–5403.

Becke, A. D., (1993). Density-functional thermochemistry: III. The role of exact exchange. *The Journal of Chemical Physics, 98*(7), 5648–5652.

Bickelhaupt, F. M., & Baerends, E. J., (2007). Kohn-sham density functional theory: Predicting and understanding chemistry. *Reviews in Computational Chemistry, 15*, 1–86.

Bubnova, O., Khan, Z. U., Wang, H., Braun, S., Evans, D. R., Fabretto, M., Hojati-Talemi, P., et al., (2014). Semi-metallic polymers. *Nat Mater, 13*(2), 190–194.

Burke, K., (2012). Perspective on density functional theory. *The Journal of Chemical Physics, 136*(15), 150901.

Capelle, K., (2006). A bird's-eye view of density-functional theory. *Brazilian Journal of Physics, 36*(4A), 1318–1343.

Car, R., & Parrinello, M., (1985). Unified approach for molecular dynamics and density-functional theory. *Physical Review Letters, 55*(22), 2471.

Cohen, A. J., Mori-Sánchez, P., & Yang, W., (2012). Challenges for density functional theory. *Chemical Reviews, 112*(1), 289–320.

Corminboeuf, C., Tran, F., & Weber, J., (2006). The role of density functional theory in chemistry: Some historical landmarks and applications to zeolites. *Journal of Molecular Structure: Theochem., 762*(1), 1–7.

Cremer, D., (1998). Møller-Plesset perturbation theory. *Encyclopedia of Computational Chemistry*.

Frisch, M. J., Trucks, M. J., Schlegel, H. B., Scuseria, G. E., Robb, M. A., Cheeseman, J. R., Scalmani, G., et al., (2009). *Gaussian, 09.*

Gaydaenko, V., & Nikulin, V., (1970). Born-Mayer interatomic potential for atoms with Z= 2 to Z= 36. *Chemical Physics Letters, 7*(3), 360–362.

Geerlings, P., De Proft, F., & Langenaeker, W., (2003). Conceptual density functional theory. *Chemical Reviews, 103*(5), 1793–1874.

Gierschner, J., Cornil, J., & Egelhaaf, H. J., (2007). Optical band gaps of π-conjugated organic materials at the polymer limit: Experiment and theory. *Advanced Materials, 19*(2), 173–191.

Gordon, R. G., & Kim, Y. S., (1972). Theory for the forces between closed-shell atoms and molecules. *The Journal of Chemical Physics, 56*(6), 3122–3133.

Gräfenstein, J., & Cremer, D., (2009). An efficient algorithm for the density-functional theory treatment of dispersion interactions. *The Journal of Chemical Physics, 130*(12), 124105.

Grimme, S., (2006). Semi empirical GGA-type density functional constructed with a long-range dispersion correction. *Journal of Computational Chemistry, 27*(15), 1787–1799.

Grimme, S., Antony, J., Ehrlich, S., & Krieg, H., (2010). A consistent and accurate Ab initio parameterization of density functional dispersion correction (DFT-D) for the 94 elements H-Pu. *The Journal of Chemical Physics, 132*(15), 154104.

Gunnarsson, O., & Lundqvist, B. I., (1976). Exchange and correlation in atoms, molecules, and solids by the spin-density-functional formalism. *Physical Review B, 13*(10), 4274.

Heeger, A. J., (2010). Semiconducting polymers: The third generation. *Chemical Society Reviews, 39*(7), 2354–2371.

Hohenberg, P., & Kohn, W., (1964). Inhomogeneous electron gas. *Physical Review, 136*(3B), B864–B871.

Jacquemin, D., Mennucci, B., & Adamo, C., (2011). Excited-state calculations with TD-DFT: From benchmarks to simulations in complex environments. *Physical Chemistry Chemical Physics, 13*(38), 16987–16998.

Jain, A., Shin, Y., & Persson, K. A., (2016). *Computational Predictions of Energy Materials Using Density Functional Theory* (Vol. 1, pp. 15004).

Kohn, W., & Sham, L. J., (1965). Self-consistent equations including exchange and correlation effects. *Physical Review, 140*(4A), A1133-A1138.

Kohn, W., (1999). Nobel lecture: Electronic structure of matter-wave functions and density functionals. *Reviews of Modern Physics, 71*(5), 1253–1266.

Kryachko, E. S., & Ludeña, E. V., (2014). Density functional theory: Foundations reviewed. *Physics Reports, 544*(2), 123–239.

Larsen, R. E., (2016). Simple extrapolation method to predict the electronic structure of conjugated polymers from calculations on oligomers. *The Journal of Physical Chemistry C, 120*(18), 9650–9660.

Laurent, A. D., & Jacquemin, D., (2013). TD-DFT benchmarks: A review. *International Journal of Quantum Chemistry, 113*(17), 2019–2039.

Manne, R., & Åberg, T., (1970). Koopmans' theorem for inner-shell ionization. *Chemical Physics Letters, 7*(2), 282–284.

Müllen, K., & Wegner, G., (2008). *Electronic Materials: The Oligomer Approach.* John Wiley & Sons.

Oliveira, E. F., Roldao, J. C., Milián-Medina, B., Lavarda, F. C., & Gierschner, J., (2016). Calculation of low band gap homopolymers: Comparison of TD-DFT methods with experimental oligomer series. *Chemical Physics Letters, 645*, 169–173.

Pastore, M., Fantacci, S., & De Angelis, F., (2010). Ab initio determination of ground and excited state oxidation potentials of organic chromophores for dye-sensitized solar cells. *The Journal of Physical Chemistry C, 114*(51), 22742–22750.

Perdew, J. P., Burke, K., & Ernzerhof, M., (1996). Generalized gradient approximation made simple. *Physical Review Letters, 77*(18), 3865–3868.

Perdew, J. P., Chevary, J. A., Vosko, S. H., Jackson, K. A., Pederson, M. R., Singh, D. J., & Fiolhais, C., (1992). Atoms, molecules, solids, and surfaces: Applications of the generalized gradient approximation for exchange and correlation. *Physical Review B, 46*(11), 6671.

Peverati, R., & Truhlar, D. G., (2014). Quest for a universal density functional: The accuracy of density functionals across a broad spectrum of databases in chemistry and physics. *Phil. Trans. R. Soc. A., 372*(2011), 20120476.

Runge, E., & Gross, E. K. U., (1984). Density-functional theory for time-dependent systems. *Physical Review Letters, 52*(12), 997–1000.

Salzner, U., & Aydin, A., (2011). Improved prediction of properties of π-conjugated oligomers with range-separated hybrid density functionals. *Journal of Chemical Theory and Computation, 7*(8), 2568–2583.

Salzner, U., (2014). Electronic structure of conducting organic polymers: Insights from time-dependent density functional theory. *Wiley Interdisciplinary Reviews: Computational Molecular Science, 4*(6), 601–622.

Schrödinger, E., (1926). Quantisierung als eigenwertproblem. *Annalen Der Physik, 385*(13), 437–490.

Senn, H. M., & Thiel, W., (2009). QM/MM methods for biomolecular systems. *Angewandte Chemie International Edition, 48*(7), 1198–1229.

Stewart, J. J. P., (1989a). Optimization of parameters for semi empirical methods I. Method. *Journal of Computational Chemistry, 10*(2), 209–220.

Stewart, J. J. P., (1989b). Optimization of parameters for semi empirical methods II. Applications. *Journal of Computational Chemistry, 10*(2), 221–264.

Stewart, J. J. P., (1991). Optimization of parameters for semiempirical methods. III Extension of PM3 to Be, Mg, Zn, Ga, Ge, As, Se, Cd, In, Sn, Sb, Te, Hg, Tl, Pb, and Bi. *Journal of Computational Chemistry, 12*(3), 320–341.

Stewart, J. J. P., (2007). Optimization of parameters for semi empirical methods V: Modification of NDDO approximations and application to 70 elements. *Journal of Molecular Modeling, 13*(12), 1173–1213.

Stewart, J. J. P., (2013). Optimization of parameters for semi empirical methods VI: More modifications to the NDDO approximations and re-optimization of parameters. *Journal of Molecular Modeling, 19*(1), 1–32.

Svensson, M., et al., (1996). ONIOM: A multilayered integrated MO+MM method for geometry optimizations and single point energy predictions. A test for Diels-Alder reactions and Pt (P (t-Bu) 3) 2+H2 oxidative addition. *The Journal of Physical Chemistry, 100*(50), 19357–19363.

Te Velde, G. T., Bickelhaupt, F. M., Baerends, E. J., Fonseca, G. C., Van, G. S. J., Snijders, J. G., & Ziegler, T., (2001). Chemistry with ADF. *Journal of Computational Chemistry, 22*(9), 931–967.

Tong, B., & Sham, L., (1966). Application of a self-consistent scheme including exchange and correlation effects to atoms. *Physical Review, 144*(1), 1.

Van, L. R., (1999). Mapping from densities to potentials in time-dependent density-functional theory. *Physical Review Letters, 82*(19), 3863.

Wang, Y., & Perdew, J. P., (1991). Correlation hole of the spin-polarized electron gas, with exact small-wave-vector and high-density scaling. *Physical Review B, 44*(24), 13298–13307.

Wykes, M., Milián-Medina, B., & Gierschner, J., (2013). Computational engineering of low band gap copolymers. *Frontiers in Chemistry*, 1.

CHAPTER 4

Photoluminescence and Electroluminescence in Semiconducting Polymers

ANDREIA DE MORAIS,[1] JILIAN N. DE FREITAS,[1] DIEGO DE AZEVEDO,[2] and RAQUEL A. DOMINGUES[3]

[1]CTI – Renato Archer Information Technology Center, Dom Pedro I highway, km 143.6, 13069-901, Campinas Brazil

[2]Chemistry Institute, University of Campinas-UNICAMP, P.O. Box – 6154, CEP 13083-970, Campinas, SP, Brazil

[3]Institute of Science and Technology, Federal University of São Paulo-UNIFESP, R. Talim, 330, 12231-280, São José dos Campos, SP, Brazil

4.1 INTRODUCTION

In semiconducting polymers (SemPolys), the electrons forming the double bonds are delocalized over the main chain. This conjugated system can be disturbed by numerous factors, including conformational defects or the presence of chemical impurities in the material, thus leading to interruptions in the conjugation. Consequently, the polymeric chains possess a collection of conjugated segments of different lengths. Such segments determine the optical and electronic properties of the polymers and are usually referred to as "chromophores."

The absorption of energy by the chain segments results in the formation of excited states, producing excitons. The excitons generated in polymer chromophores share similarities with the excited states of small molecules, so that their optical properties may be described according to a molecular approach. Nevertheless, in a polymer thin film, it is crucial to consider that there is a distribution of chromophores which are in close contact and, therefore, interact with each other. When the excited states decay with the emission of visible light, the emission is called luminescence. The luminescence may be a result of different types of excitation: when the electronic excitation occurs due to the incidence of non-ionizing electromagnetic radiation (photons), the process of light emission is denominated photoluminescence (PL), whereas the electroluminescence (EL) happens when the passage of an electric current generates the excited state through the material (via recombination of injected charges).

Overall, the decay processes are considered similar when the deactivation of the excited state occurs via the Franck-Condon radiative relaxation (which believes that the nucleus does not move during an electronic transition); regardless of the exciton formation process (Lakowicz, 1999). Thus, the most different point between the PL and EL phenomena is related to the pathway of formation of the excited states. While in the PL the excitation originates from the incidence of photons in the material, the EL requires the injection of electrons and holes from opposite electrodes, the capture of one type of charge carrier by the other (recombination) and, as a result from this recombination process, the radioactive decay of the excited state (Friend et al., 1999). Accordingly, the observation of the EL phenomenon is intimately related to the application of a voltage bias, which requires the use of electrodes and the formation of metal-semiconductor interfaces. One of the earliest observations of the PL phenomenon in organic compounds was reported in 1845, for quinine molecules exposed to solar radiation (Lakowicz, 1999). The first view of the EL phenomenon in organic molecules was reported in 1960, for anthracene crystals (Friend et al., 1999), and SemPolys, it was first published in 1990 (Burroughes et al., 1990).

Because of the need of applying a voltage to generate a current flow through the device, the EL phenomena cannot occur in an isolated material, i.e., it must be associated with interfaces (electrodes), from where the injection of charge carriers can happen (Zakya, 2005). The most straightforward device architecture necessary to observe the EL process is a single layer organic light-emitting diode (OLED). Overall, it consists of a transparent substrate (glass or an inert polymer) onto which is deposited a thin, transparent layer of an anode constituted of a high ionization potential metal alloy that is responsible for the injection of positive charges. On top of the anode, the light-emitting organic material is deposited, generally with a thickness varying between 30–300 nm. This layer is referred to as the active layer. To achieve an efficient light-emitting process, it is vital that this emissive layer has a uniform thickness and low surface roughness. On the surface of the active layer, a metal layer is thermally evaporated to form the cathode. The cathode must have a low work function since it is responsible for the injection of electrons in the active layer. The cathode and anode work functions must be compatible with the energy levels of the electroluminescent material used in the active layer. Usually, ITO (indium tin oxide) is used as a hole injector electrode, while Al, Mg, or Ca are selected for electron injection.

The operation of an electroluminescent device thus follows four steps: (i) injection of charge carriers from the electrodes into the active layer; (ii) charge carrier transport; (iii) recombination of electrons and holes into excitons and (iv) exciton migration and radiative decay, producing the light emission. Upon the application of an electric field (voltage), the cathode injects electrons into the conduction band of the electroluminescent material (orbitals π^*, or LUMO), and the anode injects holes into the valence band of the electroluminescent material (orbitals π, or HOMO). The electrons and holes flow through the emitting layer in opposite directions and, when interacting coulombically, recombine forming the exciton. The radiative decay of the exciton gives rise to the EL of the device. In the radiative process, the released photon has energy characterized by the energy gap between the LUMO and the HOMO levels of the chromophore. The color of the emitted light is controlled by the bandgap energy (E_g), whereas the charge injection process is controlled by the energetic differences between the work functions of the respective electrodes and the electron affinity (E_a) (cathode injection) and ionization potential (I) (anode injection) of the polymer.

The capture process (electron-hole recombination) in these devices is also crucial for proper operation, and to obtain a device with high efficiency, it is necessary to have a right balance between the injection of electrons and holes and charge carrier mobility, so that there will be no positive or negative charges in excess traveling through the device. An imbalance can lead to luminescence extinction. Nevertheless, the local charge density must be sufficiently high to ensure that charge carriers oppositely charged will pass within the collision capture radius.

In the EL process, the electron-hole capture is generally a spin-independent process, whereas, in the PL process, the occurrence of singlet or triplet excited states follows the spin multiplicity rule, as will be detailed in Section 4.2.

4.2   MULTIPLICITY OF THE EXCITED STATES AND QUANTUM EFFICIENCY

Excitation of a sample with electromagnetic radiation is achieved when a light beam irradiates it. Photon absorption by a conjugated polymer can lead to excitation of a specific chemical bond, chemical group, or a chain segment, depending on the excitation energy. There must be a resonance between the energy of the incident photon and the energy levels involved in the transition (Birks, 1970). To each electronic state, there are vibrational and rotational states associated. In PL or EL, the light emission (luminescence) can happen as fluorescence, involving transitions with the same spin multiplicity, or as phosphorescence, which occurs from the first triplet excited state to the fundamental singlet state.

The probability of transition is limited by selection rules that arise from characteristics of the excited states, such as the energetic difference, spin multiplicity, and symmetry. In a more detailed outlook about the spin-multiplicity selection rule, the electric dipole transitions between states with different spin multiplicity are formally forbidden by the quantum mechanical selection rule.

The transition dipole moment of an electronic transition, from *lm* electronic state to *un* state is expressed by Eq. (4.1) as follow:

$$M_{lm,un}\left(Q_{eq}^{lm}\right) = \int \ddot{}_{lm}^{*}\left(Q_{eq}^{lm}\right)\sum_{i}r_i^{\ddot{}} {}_{un}\left(Q_{eq}^{lm}\right)d\tau \qquad (4.1)$$

where, Q_{eq}^{lm} is the equilibrium nuclear configuration, at the initial electronic state *lm*, and $\ddot{}_{ln}\left(Q_{eq}^{lm}\right)$ and $\ddot{}_{un}\left(Q_{eq}^{lm}\right)$ are *lm* and *un* wavefunctions, respectively, at the *lm* equilibrium state configuration. If there is no spin-orbital coupling, these wavefunctions can be expressed separately in electronic and nuclear spatial functions. Thus, the transition moment can be expressed by Eq. (4.2) as follow:

$$M_{lm,un}\left(Q_{eq}^{lm}\right) = \int \Psi^{*}_{lm(space)}\left(Q_{eq}^{lm}\right)\sum_{i}r_i\Psi_{un(space)}\left(Q_{eq}^{lm}\right)dq\int\Psi^{*}_{lm(spin)}\Psi_{un(spin)}d\sigma$$

$$(4.2)$$

The spin functions ($\Psi^{*}_{lm\,(spin)}$ $\Psi_{un(spin)}$) are orthonormal and, if they are different from each other, their overlapping integral becomes zero and the transition moment is extinguished. However, if the spin functions are identical, the integral assumes the value 1 and the expression representing the transition moment is described by Eq. (4.3) as follow:

$$M_{\text{lm},un}\left(Q_{eq}^{\text{lm}}\right) = \int \Psi\, l^*_{m(space)}\left(Q_{eq}^{\text{lm}}\right)\sum_i r_i \Psi_{un(space)}\left(Q_{eq}^{\text{lm}}\right) dq \qquad (4.3)$$

The implication that follows is that transitions between states of the same spin multiplicity are allowed. It is important to note that this selection rule assumes that there are no spin-orbital interactions, when, in fact, electrons can change their spin very slowly due to spin-orbital and spin-spin interactions, which ultimately results in the fact that transitions between states with different spin multiplicities may occur, but at a low probability. This spin multiplicity selection rule has a great influence on the photophysical processes of aromatic molecules (Birks, 1970).

In summary, when the excitation of organic molecules is driven by non-ionizing electromagnetic radiation (photophysical process), the probability of a singlet-singlet transition is many orders of magnitude larger than a singlet-triplet transition, since the latter is a transition forbidden by the spin multiplicity selection rule. Nevertheless, the emergence of an excited triplet state can occur due to a phenomenon called intersystem crossing (ISC) (Birks, 1970).

On the other hand, in the electrical excitation process the electron-hole capture process is spin-independent (being only affected by the spin-statistics). The excitons are then formed with triplet and singlet configurations at the 3:1 ratio. The exciton spin wavefunction, formed from two electronically charged spins ½, can be either singlet or triplet, and the consequence of the excitation confinement is that the energy difference between singlet and triplet (exchange energy) must be large, being the passage from the triplet to the singlet state unlikely. The triplet excitation does not produce fluorescence emission, occurring triplet-triplet annihilation, or a phosphorescence process. Because only singlets can decay radiatively, and there is only one singlet for each three triplet states, the maximum quantum efficiency (number of photons emitted per electrons injected) that could be achieved with fluorescent polymers is, theoretically, 25%.

Despite the similarities in the PL and EL emission spectra profile, the quantum yields are usually different. PL quantum yield can be defined as the ratio between the numbers of emitted photons per number of absorbed photons, whereas in the EL, it results from the recombination of polaronic pairs and exciton decays that occur in an emissive way. It is also important to emphasize that the quantum yield can be easily determined in dilute solutions, but it is difficult to determine PL or EL quantum yields in the solid-state, either because of the absence of standards, or the complexity of solid-state energy transfer processes. Generally, the literature discusses that the luminescence efficiency of polymers is much lower in the solid-state than in solution, because the non-radiative decay rate can be influenced by the energy transfer process, which will be discussed in detail in Section 4.3.

While the PL quantum efficiency (φ_{PL}) depends basically on internal processes that occur within the molecule or in the polymer network, the EL quantum efficiency (φ_{EL}) also depends on other factors, such as the device structure, electron, and hole current balance, and the effectiveness of electron and hole injection and recombination.

Overall, φ_{PL} can be expressed with the following equation:

$$\phi_{PL} = \frac{\Gamma}{\Gamma + k_{nr}} \tag{4.4}$$

where, Γ is the emission rate and k_{nr} is the non-radiative decay rate (Lakowicz, 1999); and φ_{EL} can be defined by:

$$\phi_{EL} = \gamma r_s q \tag{4.5}$$

where, γ is the ratio between the exciton formation within the device and the number of electrons circulating in the external circuit, r_s is the ratio of formation of singlet excitons, and q is the decay efficiency of these singlet states (Friend et al., 1999). In theory, r_s should be limited to 25% if it is determined by spin-statistics, as discussed above.

In the PL, a certain number of excitons (I) are generated initially in the singlet state (S_1), and that excited state may decay via radiative or non-radiative processes to the ground state (S_0). Also, ISC to the triplet state (T_1) may also happen, and both radiative and non-radiative emissions can occur from T_1 to S_0 state. In the EL, the processes are the same, with the important exception that it is also possible to create a χ_T fraction of triplet excitons directly, since the formation of the excited states is spin-independent. A model for the PL and EL processes is illustrated in Figure 4.1 (Wilson et al., 2001).

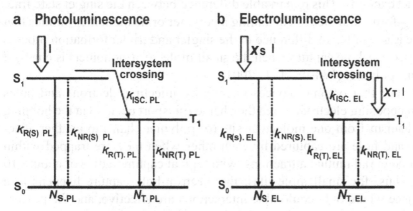

FIGURE 4.1   (a) PL and (b) EL emission models. I represent the initial number of excitons; k_{NR}, k_R, k_{ISC} represent the non-radiative, radiative, and intersystem crossing rate processes; N_s and N_T represent the number of singlet and triplet photons that decay in the emissive form; χ_S and χ_T are the fractions of singlet and triplet states generated, respectively; and S_0 is the ground state.

Source: Reprinted with permission from Wilson et al. (2001). Copyright 2001, Macmillan Publishers Ltd.

Considering these processes, the PL and EL of OLEDs containing a conjugated polymer containing Pt atoms covalently bound to the chain, or its monomer, were investigated to determine the total fraction of singlet states generated in these materials. The heavy atom induces spin-orbit coupling, favoring triplet emissive transitions (phosphorescence) so that the optically and electrically generated luminescence from both singlet and triplet states could be compared. Thus, by performing a comparison between PL and EL and associating these processes, the authors investigated the quantum yield (φ) for the triplet (TR) and singlet (SR) radiative emission and for the ISC (Wilson et al., 2001). With the model shown in Figure 4.2, the measurements of the PL and EL spectra, and if there is no difference in the decay rates for excited states in PL and EL, it was possible to determine the fraction of singlet excitons that were generated by electric excitation. The value found in the study was 54% for the diode assembled with the polymeric material, and ~22% for the monomeric material. These values suggest that, while in the monomeric material recombination is indeed dictated by the spin statistic (limited to 25%), in the polymeric system there may be spin-dependent processes, favoring the formation of singlet excitons. This remarkable difference between the singlet state fraction χ_s formed in diodes containing a polymer or its monomer implies that there is a significant difference in the singlet and triplet formation process via electron-hole capture when the small molecule (monomer) is changed for the polymer (Wilson et al., 2001).

Given that organic devices operate by injecting electrons and holes from opposing electrodes, and the charge transport occurs via the hopping mechanism from one molecular site (or polymer chain) to another, electron, and hole are captured by each other when they are trapped within their potential mutual attractions, with typical capture radius of about ~10 nm. Thus, for small molecules, the electron-hole capture happens at a distance where only coulombic interactions are effective, and the process is independent of the exciton final spin state. In polymers, on the other hand, when the electron and hole are present in the same chromophore, within the capture radius, the interaction between them also results from the direct overlapping of electron and hole wavefunctions, creating some spin dependence (Wilson et al., 2001).

Another significant parameter affecting the internal quantum efficiency of luminescence is the competition between radiative and non-radiative decays of the electron-hole pairs created within the polymer layer. These pairs can migrate along the chains and are therefore susceptible to trapping at quenching sites where non-radiative processes may occur. They may also undergo a phonon emission and loose energy in a thermal burst, or transfer energy to 'impurities,' or convert into a triplet by ISC and eventually lose energy non-radiatively. This subject will be further explored in Section 4.3.

4.3   TYPES OF EXCITED STATES AND ENERGY TRANSFER PROCESSES

The occurrence of bimolecular processes also determines the emission spectra. Five types of emissive states can be observed in conjugated polymers: excimer, exciplex, electromer, electroplex, and aggregates. The dimers excimer, exciplex, electromer, and electroplex are frequently seen in small molecules, as well as in polymeric materials. However, the aggregate, a species with a ground-state π-π interaction, has a much stronger contribution to polymers. Energy transfer from isolated chain segments to aggregates can occur and play an essential role in the PL and EL properties of conjugated polymers.

Excimers are created when an electronically excited molecule (M*), before deactivation, interacts with another identical molecule in the fundamental state (M). This type of interaction occurs at short distances (Birks, 1970). Excimers can be formed for both types of excitation: photo-excitation or electron excitation. On the other hand, the electromer can be formed only when the excitation occurs by charge recombination. This type of emissive state occurs in a direct transition from the LUMO of one molecule to the HOMO of another molecule, and so this process is also called cross-transition (Kalinowski et al., 2000). The electromer emission occurs when, because of a defect, electron transfer from LUMO to LUMO and/or hole transfer from HOMO to HOMO is hindered. The emission is then redshifted in comparison to the molecular emission, as also observed in the emission of excimers. It is important to note that the electromer emission requires the charge carriers to be previously separated, for example, by exciton dissociation (Yanga et al., 2007).

Consequently, when the formation of electromers occurs, the EL spectrum is expected to be different, or more complex, than the PL spectrum. In this case, the light absorbed by molecules of one type produces donors in the excited singlet state (D*) and these species may, through radiative relaxations, present fluorescence decay (hv_D). In another path, the D* species may interact with the electron acceptor in the ground state (A), forming an exciplex (for both PL and EL processes) or an electroplex (only EL process) (Birks, 1970; Giro et al., 2000). The exciplex is another type of dimer that behaves similarly to excimer but consists of interaction between different chromophores. The electroplex is also formed by two different molecules and are analogous to the electromers, formed when only one component is involved. The EL spectrum of these molecular systems may show many bands, corresponding to the different light-emitting species. Because of these features, the EL spectra can be broader, with components shifts to lower energy regions, in comparison to the PL spectra. Figure 4.2 presents a scheme of the formation of excited states in electroluminescent diodes (Kalinowski, 2008).

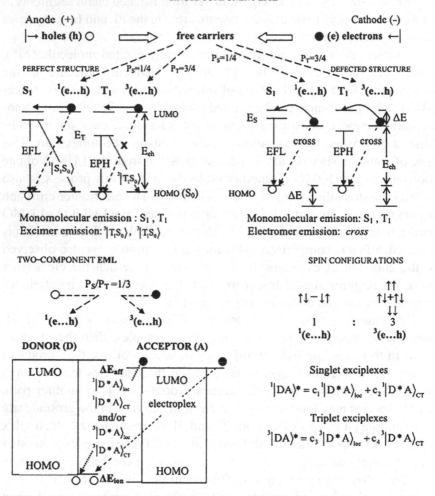

FIGURE 4.2 Scheme illustrating the formation of bimolecular excited states by electron-hole recombination in the emissive layer (EML) of electroluminescent diodes.

Source: Reprinted with permission from Kalinowski (2008). Copyright 2008, Elsevier.

The formation of aggregates involves the interaction between two or more chromophores in the ground state, extending the delocalization of π-electrons in these aggregated segments, thus leading to the generation of spectral responses in the longer wavelength spectral region. This process, sometimes, is associated with a decrease of the luminescence quantum efficiency. Aggregates can be observed in both PL and EL spectra. Generally, non-interacting isolated chain and aggregates with various extents of intrachain and interchain interactions coexist in the solid-state and concentrated solutions. In intrachain interactions, excitons can be formed upon optical irradiation, and their excitation energy may be released via non-radiative/radiative relaxations or by energy transfer processes to other species of lower energies. On the other hand, due to the packing of polymer chains, polymeric films are susceptible to interchain excitations, after which excitons can re-form to result in delayed luminescence, which can be redshifted in single-chain emission. The fluorescence color can be drastically redshifted, and the light evolved from concentrated solutions and films may even be entirely from the aggregation emission.

The energy transfer between chromophores can occur both in solution and in solid-state, affecting the PL or EL, and is classified into two distinct processes: radiative and non-radiative. The radiative process, also called "trivial," is a two-step process in which a chromophore (D) emits a photon and another molecular species (A) absorbs this radiation, as described in the following equation:

$$D^* \rightarrow D + h\nu$$

$$h\nu + A \rightarrow A^*$$

(4.6)

The trivial mechanism does not require energetic interaction between the donor and the acceptor, and its efficiency depends on the spectral overlap between the emission spectra of the donor and the absorption spectra of the acceptor (Birks, 1970). This process can occur between identical molecules, which may lead to self-absorption or re-emission processes, as well as the internal filter effect, in both concentrated solutions and solids. In this kind of energy transfer, there is a decrease in the intensity of the emission of light with the wavelengths corresponding to the spectral overlap region.

In the trivial process, the molecule D^* can be excited either by a photoluminescent or an electroluminescent process. However, the formation of A^* is necessarily a photoluminescent process, as it occurs through the absorption of photons emitted during the relaxation of the D^* species. Thus, if the trivial mechanism exists in an electroluminescent device, PL emission may occur after an EL emission.

Another energy transfer process is the non-radiative via Förster mechanism (*Fluorescence Resonance Energy Transfer,* FRET) which occurs in a single step, involving a dipole-dipole coupling between the donor (D) and the acceptor (A), as described by the following equation:

$$D^* + A \rightarrow D + A^* \tag{4.7}$$

The probability of energy transfer via non-radiative mechanism is dictated by the distance between the donor and the acceptor, the relative orientation between them, their spectroscopic properties, and the optical properties of the surrounding environment, among other factors (Birks, 1970). Its efficiency also depends on the spectral overlap between the emission spectra of the donor and the absorption spectra of the acceptor. A reduction of the intensity of the emission and the lifetime of the excited state of the donor species (D*) are evidence for the occurrence of this mechanism (Lakowicz, 1999). This process causes changes in the spectrum profile since the D* species decays without light emission.

In a polymer film (solid-state), the optoelectronic properties are determined by emitting species arising from intrachain or interchain interactions. Such interactions inevitably change the emitting color and may, sometimes, the emission intensity. The emission spectra are usually redshifted due to aggregation effects, internal filter, self-absorption, and re-emission processes. These effects may reduce the PL or EL intensity since the possibilities of non-radiative deactivation or unwanted emissions are higher in more concentrated samples. This phenomenon may quench or decrease the intensity of the higher energy emission band, and when the energy transfer process is very efficient, the higher energy component may be completely absent in the emission spectra.

Nevertheless, this effect can also be explored to enhance the emission properties of polymeric OLEDs. When a reasonable concentration of the lower energy component is used, both color tuning and an increase in the luminous performance can be achieved simultaneously (Quites, 2014). The blending of polymer materials, for example, can be used to tune the light emission characteristics, with simultaneous luminescence enhancement. The use of the poly(9,9-di-n-octylfluorenyl-2,7-diyl), with relatively large bandgap function, as the host material, combined with low amounts of the green light-emitting copolymer poly[(9,9-di-n-octylfluorenyl-2,7-diyl)-alt-(benzo[2,1,3]thiadiazol-4,8-diyl)] has been used to enhance the brightness of the OLEDs (Buckley et al., 2001; Chappell et al., 2003; Jokinen et al., 2015; Pereira et al., 2016). Polymer blending leads to higher emission yields due to the dilution effect, which decreases exciton-exciton annihilation and Förster energy transfer from the wider bandgap material (donor) to the narrower bandgap component (acceptor). The excitons created upon absorption of the higher energy photons at donor sites transfer the energy to acceptor molecules, reducing the degree of exciton quenching and the excimer emission, and enhancing the PL or EL efficiency of the acceptor.

As discussed in this section, the PL and EL spectra of a polymer, or polymer blends, can present different features, as a result from the existence of non-interacting isolated chain segments and aggregates with various extents of intrachain and interchain interactions. Thus, relevant information can be driven by the analysis of the fluorescent behavior of polymers, including the assessment of the organization/structure of the polymeric film. In Section 4.4, some techniques usually employed to investigate the PL and EL of polymers will be addressed.

4.4 TECHNIQUES FOR ANALYSIS OF THE PL AND EL PROCESSES

Since its inception, optical spectroscopy has steadily expanded to an amazing variety of fields of material science. The technique itself also has experienced a spectacular breakthrough with the emergence of high-resolution microscopes, which allows imaging in real-time on nanometer scales (Orrit et al., 2014). In this section, we will highlight some articles to illustrate the power and versatility of the optical spectroscopy and fluorescence microscopy techniques for the study of the luminescence of conjugated polymers.

The two most successful and widely used fluorescence microscopy modes to detect single-molecules and aggregates in conjugated polymers are confocal scanning microscopy and wide-field fluorescence microscopy (Moerner et al., 2003). Figure 4.3(a) shows the essential components of a confocal scanning microscope such as an objective lens, a dichroic mirror, filters, pinhole, and single-photon counting module (SPCM) (Chen et al., 2014). This method, also called confocal laser scanning microscopy (CLSM), consists of the excitation light from the laser that passes through a waveplate before being reflected towards the sample by a dichroic mirror (White et al., 2002). Waveplate is an optical device that alters the polarization state of a light wave traveling through it. The objective lens focuses the excitation light to a limited diffraction spot. The PL emitted by single-molecules or aggregates is collected by the objective lens, and after passing through the dichroic mirror, is filtered by appropriate filters and a pinhole before reaching the photodetector to reject the background fluorescence and out of focus light, respectively (White et al., 2002; Wöll et al., 2009; Chen et al., 2014). Large area single-molecules and aggregates images are obtained point-by-point by scanning the sample, and the fluorescence intensity of individual molecules is recorded one-by-one (Chen et al., 2014).

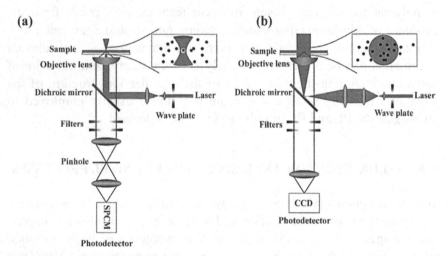

FIGURE 4.3 (a) Confocal scanning microscopy setup, (b) wide-field fluorescence microscopy setup.

Source: Reprinted with permission from Chen et al. (2014). Copyright 2014, Molecular Diversity Preservation International and Multidisciplinary Digital Publishing Institute (MDPI).

The confocal scanning microscopy was used to identify the positions of single-molecules and aggregates formed in poly(*p*-phenylene-ethynylene-butadiynylene) (PPEB) / poly(methyl-methacrylate) (PMMA) polymer films. Figure 4.4(b–e) shows a series of confocal images for PPEB/PMMA films with different solvent mixtures used for solvent vapor annealing (SVA). The acetone: chloroform ratios were 100:0, 90:10, and 80:20. In this way, the authors observed the formation of ordered multichain aggregates with a predetermined average size in each experimental condition. Green denotes the PL detected below 532 nm, originating from isolated chains, and red, the PL above 532 nm from aggregates. After SVA with a vapor of 80:20 acetone: chloroform ratio (Figure 4.4(e)), the overall number of spots decreased whereas the brightness of the red spots considerably increased. The fraction of red PL, F_{red}, was also calculated as I_{red} / (I_{red} + I_{green}) (Figure 4.4(a)) and shown for each particle in the histograms (Figure 4.4(b–c)) (Stangl et al., 2015).

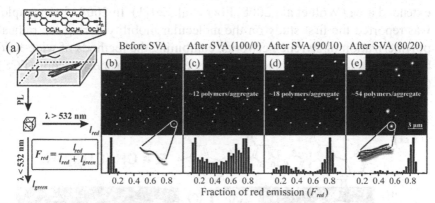

FIGURE 4.4 (a) Schematic illustration of the splitting of single-spot PL into two detection channels, for photons with λ >532 nm (denoted I_{red}) and λ < 532 nm (denoted I_{green}). The fraction of red emission for every single spot is defined as F_{red}. (b–c) Confocal scanning microscope images of the controlled growth of conjugated polymer aggregates from isolated single chains under different processing conditions. The corresponding F_{red} values are shown in histograms below each image.

Source: Reprinted with permission from Stangl et al. (2015). Copyright 2015, Proceedings of the National Academy of Sciences (PNAS) of the United States of America.

In a recent work of Steiner et al. (2017) the aggregation process by SVA was induced in a sample containing high poly(3-hexylthiophene) (P3HT) concentration. The structure of this conjugated polymer is illustrated in Figure 4.5(a). In that work, SVA was performed for 30 min on the microscope setup using a gas-flow chamber with a mixture of 95% acetone and 5% chloroform vapor. The confocal fluorescence scan images were obtained for a sample before and after SVA, as shown in Figure 4.5(b) and 4.5(c), respectively.

In general, the limitation of the use of confocal scanning microscopy is that only one single-molecule or aggregate can be detected at a time, which increases the detection time if a larger area is investigated (Chen et al., 2014). Perhaps the simplest method to observe single-molecule or fluorescence aggregates is to use wide-field microscopy (see Figure 4.3(b)). Unlike a confocal scanning microscope, it is possible to obtain images of several individual molecules simultaneously in a wide-field fluorescence microscope, leading to a loss of resolution in the image. The large area images are directly recorded by a charge-coupled diode (CCD) (Wöll et al., 2009; Chen et al., 2014). This microscope is most commonly used to study diffusion, and other dynamic processes are occurring over an extended area (Wöll et al., 2008; Flier et al., 2011). In 2008, for example, was reported the first study on the molecular mobility during the radical polymerization of styrene networks, combining the methods of wide-field fluorescence microscopy and fluorescence correlation spectroscopy (FCS) (Wöll et al., 2008).

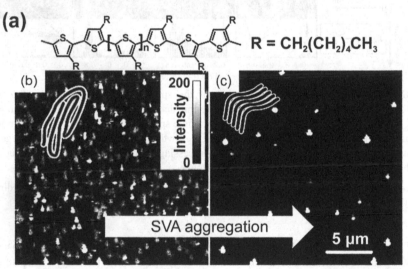

FIGURE 4.5 (a) Chemical structure of P3HT. Confocal scanning microscope images of P3HT embedded in PMMA (b) before and (c) after solvent vapor annealing (SVA).

Source: Reprinted with permission from Steiner et al. (2017). Copyright 2017, American Chemical Society.

For conjugated polymers, it was recently used the wide-field fluorescence microscopy to demonstrate the bottom-up growth of morphologically ordered anisotropic aggregates prepared via SVA from single poly(2-methoxy-5-(2-ethylhexyloxy)-1,4-phenylenevinylene)   (MEH-PPV) chains in a host PMMA matrix. The control of some parameters, such as the choice of solvent, pressure, and flow of solvent vapor, as well as the degree and time of film swelling, have a substantial impact on the final morphology and properties of the polymer thin film. In that work, the authors observed the formation of a fiber network, which is morphologically like those present in bulk heterojunction active layers of polymer-based devices (see Figure 4.6(a–d)). They suggested that the interconnected networks of fiber aggregates may provide efficient charge transport in the active layer in these devices (Yang et al., 2017).

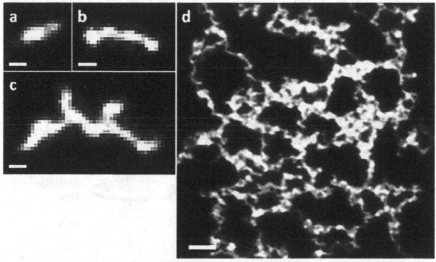

FIGURE 4.6   Wide-field fluorescence images of (a–c) MEH-PPV fiber aggregates and (d) a network thereof. Scale bars in panels (a–c) and (d) are 0.5 and 2 µm, respectively.

Source: Reprinted with permission from Yang et al. (2017). Copyright 2017, American Chemical Society.

FCS is one of the single-molecule-sensitive techniques that allow real-time access to many molecular parameters such as diffusion coefficients, concentration, and molecular interactions (Haustein and Schwille, 2004). This spectroscopy technique requires small sample amounts and low concentrations, so it allows measurements with minor disturbance of the systems, with a good spatial resolution at the diffraction limit. Most FCS studies have been conducted in biological systems. Nowadays, some works use this technique to explore polymer dynamics at interfaces, in solutions, gels, melts, and glasses, or in polymerization, micellization, and aggregation processes. An excellent review of the use of FCS in polymer systems was published in 2014 (Wöll, 2014). Typical FCS setups have also been described in various reviews (Haustein and Schwille, 2007; Koynov and Butt, 2012; Papadakis et al., 2014; Wöll, 2014). Figure 4.7 shows a summary of the basic principles of this spectroscopy technique.

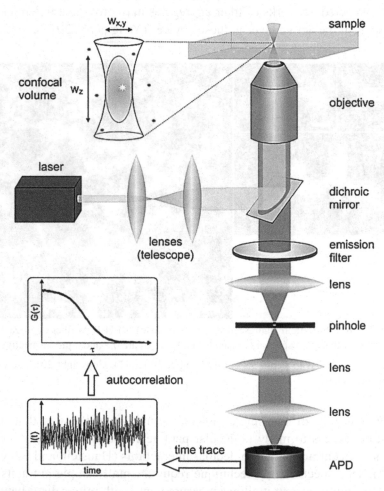

FIGURE 4.7   Typical fluorescence correlation spectroscopy (FCS) setup.

Source: Reprinted with permission from Wöll (2014). Copyright 2014, Royal Society of Chemistry

A typical setup for FCS (Figure 4.7) closely resembles that of a standard confocal scanning microscope (Figure 4.3(a)). In this case, the photons passing through the pinhole are detected with a fast and sensitive photodetector, typically a single photon counting avalanche photodiode (APD). The temporal fluctuations in the detected fluorescence intensity, $I(t)$, caused by fluorescent species diffusing through the observation volume, are recorded and analyzed by an autocorrelation function as described below (Koynov and Butt, 2012):

$$G(\tau) = \frac{\langle \delta I(t).\delta I(t+\tau) \rangle}{< \delta I(t) >^2} \quad (4.8)$$

where, $\delta I(t) = I(t) - < I(t)>$ and $<>$ denotes the time average.

The autocorrelation function ($G(\tau)$) contains information on all processes which cause intensity fluctuations within the confocal volume (e.g., translational or rotational diffusion). The intensity fluctuations in these processes arise from the change of photophysical properties of the chromophore or of its orientation (Koynov and Butt, 2012). The time axis of the autocorrelation function is often represented in a logarithmic scale, as shown in Figure 4.8.

FCS has been increasingly used in polymer science in the last years (Papadakis et al., 2014; Wöll, 2014). In recent work, it was proposed the use of FCS to investigate the aggregation process of conjugated polymers, such as P3HT, MEH-PPV, and poly[2-methoxy-5-(2-ethylhexyloxy)-1,4-phenylenevinylene] (PPV), in a common organic solvent (toluene). The authors observed for all compounds that the value of G(0) decreases after filtering, indicating that the average number of emitter species in the focal volume increases. Figure 4.8(a, b) shows an example of the FCS results of two shorter chain oligomers (OPPV7 and OPPV13), filtered, and unfiltered, in toluene. The overall shape of the normalized FCS curves was unchanged, which suggests that the average size of the emitting species was unchanged by filtration (see Figure 4.8(c)). The effect of filter size was also investigated in this work. As one example, OPPV13 in toluene, the decrease in G(0) was more substantial when the 0.1 μm versus the 0.2 μm pore filter was used (see Figure 4.8(d)) because more aggregates were disrupted. In short, it was reported that reducing aggregation by filtering the solutions before casting enhances the uniformity and emission yield of the resulting polymer films (Wu et al., 2017).

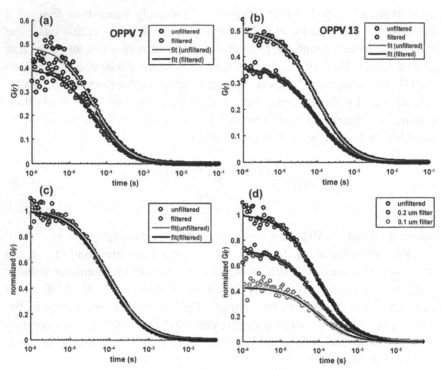

FIGURE 4.8 FCS of filtered and unfiltered polymers in toluene: (a) OPPV7 and (b) OPPV13, (c) normalized FCS curves for unfiltered and filtered OPPV13 in toluene, (d) FCS of OPPV13 in toluene using different filters.

Source: Reprinted with permission from Wu et al. (2017). Copyright 2017, American Chemical Society.

Single-molecule fluorescence spectroscopy (SMFS) has also been considered an essential tool for structural and dynamic characterization of fluorescent materials and biological systems on nanometer scales. This technique relies on the detection and measurement of the fluorescence signal and spectra from single isolated molecules or polymer chains. A typical setup for SMFS closely resembles that of an FCS (Figure 4.7) (Vacha and Habuchi, 2010). SMFS can provide information on designing electroluminescent conjugated polymer from the viewpoint of molecular scale (Chen et al., 2010).

The SMFS experiments require a highly diluted sample, where the polymer chains, molecules, or particles are well separated, spectroscopically, or spatially. The investigated material can either be immobilized in a non-emissive matrix (e.g., poly(vinyl)alcohol, PVA) or be diffusing in a solution or melt (Wöll et al., 2009). To detect fluorescence from a single molecule, such a molecule also must be sufficiently emissive, and its fluorescence intensity must be well above the background (Kondo et al., 2017). The measurements can be done as a function of time, temperature, the polarization of incident, or emitted light (Ostroverkhova et al., 2016).

Barbara and co-workers have pioneered the field of SMFS of conjugated polymers, especially for poly(2-methoxy-5-(20-ethylhexyloxy)-1,4-phenylenevinylene) (MEH-PPV) (Hu et al., 1999). The same research group published an excellent review of this subject in 2005 (Barbara et al., 2005). Others SemPolys, such as P3HT (Steiner et al., 2014; Thiessen et al., 2013) and PFO (Adachi et al., 2014; Honmou et al., 2014), have also been subjected to SMFS studies.

Among the numerous applications, SMFS can serve as a useful tool for assessing the effect of solvent/host polymer quality on the conformation of single conjugated polymer chains. The preparation of polymer chains from solvents/host polymers with different polarity can lead to notable changes in the spectra. Figure 4.9(a) shows representative intensity transients for single chains of the MEH-PPV cast from the two different solvents: toluene (a poor solvent) (top) and chloroform (a good solvent) (bottom) (Huser et al., 2000).

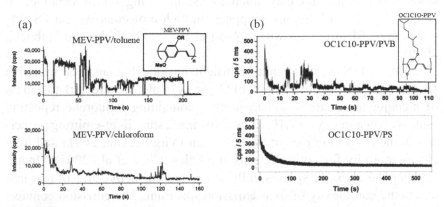

FIGURE 4.9 Comparison of fluorescence time traces of conjugated polymers in different conformational states. (a) MEH-PPV cast from the poor solvent of toluene (top) and good solvent of chloroform (bottom). *Source:* Reprinted with permission from Huser et al. (2000). Copyright 2000, Proceedings of the National Academy of Sciences (PNAS) of the United States of America. (b) OC1C10-PPV in thin-film matrix of poly(vinyl butyral) (PVB) (polar matrix) (top) and of low-molecular-weight polystyrene (PS) ($M_w = 4500$) (a polar matrix) (bottom).

Source: Reprinted with permission from Sartori et al. (2003). Copyright 2003, American Chemical Society.

In MEH-PPV/toluene samples, the PL intensity switches rapidly between on/off states, while different discrete intensity levels are assumed. This behavior can be attributed to the creation of photogenerated defects that quench excitations along the entire chain (Hu et al., 2000). In MEH-PPV/chloroform samples, the PL intensity decays gradually without discrete jumps to the background level, and on much longer timescales than that for MEH-PPV/toluene samples (Huser et al., 2000; Huser and Yan, 2001).

In another example, (Sartori et al., 2003) also used the SMFS to investigate the effect of optically inert polymers (so-called host polymers) on the spectroscopic and conformational properties of individual OC1C10-PPV (poly(2-methoxy-5-(2',6'-dimethyloctyloxy)-p-phenylenevinylene)) polymer chains (see Figure 4.9(b), inset). In that work, two polymers with different polarity were selected as host: poly(vinylbutyral-co-vinyl alcohol-co-vinyl acetate) (PVB) as the polar matrix (top) and low-molecular-weight polystyrene (PS) (M_w = 4500) as the apolar matrix (bottom). According to Figure 4.9(b), in PS (M_w = 4500) matrix, the fluorescence intensity changes were continuous, in contrast to the case of the more polar PVB matrix. However, matrix polarity is not the only important parameter, since fluorescence intensity fluctuations, including on/off behavior of single OC1C10-PPV chains, reappears in high-molecular-weight PS (M_w = 240 000) (data not shown) (Sartori et al., 2003; Vacha and Habuchi, 2010).

Also, SMFS provides unique insight into the fundamental emissive species of conjugated polymers. PFO, for example, are a remarkable class of π-conjugated materials characterized containing a fluorene repeating unit, as shown in Figure 4.10, They show interesting light-emitting properties in the deep-blue part of the spectrum (Virgili et al., 2001) and high charge mobility for both electrons and holes (Chua et al., 2005). Understanding the emissive species in PFOs has been an active area of research due to the complexity of their emission spectrum. The emission contains three components due to the presence of two distinct conformational phases (the glassy and the more ordered β-phase), as well as highly localized oxidative emissive defects (fluorenone defect). These three species contribute to ensemble emission (Becker and Lupton, 2005; Como et al., 2008). In particular, the β-phase is characterized by an extended planarization of the fluorene repeating units, while the glassy phase is a twisted chain arrangement with shorter conjugation length. These two phases have notably different optical properties in terms of luminescence, absorption, vibroniccoupling, and photophysical stability. Figure 4.10 reports two spectral traces recorded as a function of time for a β-phase and a glassy phase single PF chain (Becker and Lupton, 2005).

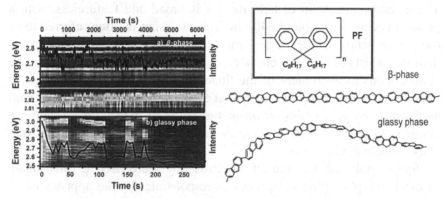

FIGURE 4.10   Temporal evolution of the PL for a β phase (a) and a glassy phase (b) PF molecule at 5 K. The integrated intensity is overlaid as a black curve. A close-up of the zero-phonon line of the β-phase PF is shown at the bottom of (a), which displays spectral diffusion.

Source: Reprinted with permission from Becker and Lupton (2005). Copyright 2005, American Chemical Society.

The planar and extended chain found in the β-phase results in a different photophysical behavior compared to glassy-phase chains. According to the Figure 4.10, the planarization results in increased stability in the emission intensity (overlaid black curve) as well as in the spectral diffusion (note the temporal scale extending up to 2 h). In contrast, the glassy phase shows a larger spectral diffusion accompanied by a short-timescale blinking and irreversible photobleaching after 200 s. In addition to this behavior, some specific photophysical properties, such as the lower extent of triplet exciton formation (Hayer et al., 2005) and efficient energy transfer from the amorphous phase to the β-phase (Khan et al., 2004) has also been reported in recent years.

The well-known keto defect in polyfluorene was used as a model to demonstrate how SMFS can directly track the formation of charge and exciton traps in conjugated polymers in real-time (Adachi et al., 2014). Many studies using SMFS have provided evidence that keto defects act as traps, which can perturb color purity in OLEDs (List et al., 2002) or promote recombination losses of photogenerated carriers in organic photovoltaics (Reid et al., 2010). The traps in polyfluorene have been determined to originate from fluorenone units formed by keto defects as a result of thermo-, photo-, or electro-oxidation (Como et al., 2008). The emission spectrum of keto defects is broad and featureless, with a peak wavelength of around 540 nm. This emission is often referred to as the "green band." Evidence was shown that it is not only broadband but instead consists of multiple emitters, each at a discrete wavelength. In the simultaneous measurement of the fluorescence lifetime and spectrum, it was observed the sequential formation of one defect after another, as well as the difference in emission properties between defects, even within a single PFO chain, suggesting substantial energetic heterogeneity of trap sites (Adachi et al., 2014).

Single-molecule EL and PL spectroscopy were used to investigate in detail the photophysical processes responsible for the appearance of the "green band" in polyfluorene chains (poly[9,9-bis(3,6-dioxaheptyl)-fluorene-2,7-diyl], BDOH-PF) confined in vertical cylinders of a phase-separated block copolymer (poly(ethylene oxide), PEO). The authors observed notable differences in the spectral properties between EL and PL and illustrated a scheme of the proposed energy levels and the optical transitions (Figure 4.11). The main difference of EL compared with PL is the absence of the emission of a singlet exciton of an isolated polyfluorene chain (SE type) and the considerable contribution of the type I spectra. In PL, only the long-wavelength type II is present. Based on quantum chemical calculations, type I was assigned to the emission of a ground-state dimer, which is formed and stabilized by a charge-assisted mechanism in the current-carrying OLED device (Honmu et al., 2014).

Other relevant properties of conjugated polymers can also be probed using SMFS, such as exciton localization/delocalization characteristics (Como et al., 2011) and intra- and intermolecular interactions (Yoo et al., 2012). Comprehensive reviews of SMFS studies promote an opportunity to understand the functional characteristics of conjugated polymers (Wöll et al., 2009; Lupton, 2010; Bolinger et al., 2012; Vogelsang and Lupton, 2012).

In optoelectronics, devices typically prepared with SemPolys, such as organic solar cells, organic field-effect transistors (FET), and OLEDs, the excitons and charge carrier's dynamics in π-conjugated polymers are essential factors for determining the performance of these devices and have been the subject of intensive research in recent years. Transient absorption spectroscopy (TAS) has also been considered a powerful tool for observing these transient species generated by photoexcitation (Ohkita and Ito, 2011; Ohkita et al., 2016). In this technique, a dye laser pumped by a nitrogen laser is usually employed as a light source for sample excitation (pump pulse), and a tungsten (W) or xenon (Xe) lamp are usually employed as a continuous light source (probe pulse) for the transmittance measurement. Furthermore, two monochromators and appropriate optical cut-off filters are placed before and after the sample to reduce unnecessary scattering light, stray light, and emission from the sample. The probe light passing through the sample is detected with a PIN photodiode. The signal is pre-amplified and sent to the main amplification system with cut-off filters to improve the signal to noise ratio. The amplified signal is collected with a digital oscilloscope, which is synchronized with a trigger signal of the laser pulse from a photodiode (Ohkita and Ito, 2011; Ohkita et al., 2016).

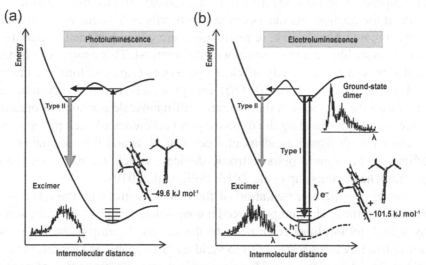

FIGURE 4.11 (a) PL and (b) EL. In the EL scheme, the neutral ground-state dimer is stabilized by the injection of a charge (for example, a hole h^+). Recombination with the opposite charge (an electron e^-) leads to the population of a neutral dimer excited state in a configuration corresponding initially to the mixed dimer ground state before recombination.

Source: Reprinted with permission from Honmu et al. (2014). Copyright 2014, Macmillan Publishers Ltd.

This type of spectroscopy is also commonly known as pump-probe spectroscopy, and it has been widely employed to detect ultrafast phenomena on a time scale of < 1 ns. Recent advances in ultra-fast techniques using lasers have allowed femtosecond-scale experiments (Kee, 2014). Exciton dynamics and formation mechanism (Ghosh et al., 2017), triplet state decay dynamics (Ohkita et al., 2006), charge transfer at the interface (Yamamoto et al., 2012), charge dissociation into free charge carriers (Dimitrov et al., 2012) and photoinduced charge generation (Kaake et al., 2012) are some of the most studied ultrafast phenomena in conjugated polymers. These processes range over a broad temporal scale, from femtoseconds to microseconds, which can be directly observed with TAS.

Pump-probe confocal microscopy is also another powerful technique widely used to investigate excited state dynamics in conjugated polymers. This technique combines the spatial resolution (~ 300 nm) of far-field optical microscopy with the temporal resolution (~ 150 fs) of ultrafast pump-probe spectroscopy (Polli et al., 2010). It has the potential to unravel the phenomena that occur at the interfaces between two materials (e.g., polymer blends). These phenomena are of fundamental importance but are incredibly complex and poorly understood. This complexity is due to the presence of a variety of electronic states (e.g., excitons, excimers, and charge transfer states (CTS)) and processes (e.g., energy transfer and charge separation) in the formation of nanoscale domains in organic materials. Understanding the dynamics and efficiency of such phenomena is crucial, both from a fundamental point of view and for optimizing the efficiency of organic optoelectronic devices, which are based on these phenomena (Grumstrup et al., 2015; Polli et al., 2010).

Generally, the representation of the results is given by the $\Delta T/T$ as a function of time. The pump-induced transmission change (ΔT) is measured by a lock-in amplifier referenced to the chopped pump-pulse train and normalized by transmission (T) to yield $\Delta T/T$ (Cabanillas-Gonzalez et al., 2011). This technique provides new insight into the properties of conjugated polymer blends by directly accessing the dynamics at the interfaces between different materials (Polli et al., 2010). In 2012, this tool was used to study the photophysics of phase-separated polymer blends: a blue-emitting PFO in an inert matrix of PMMA and an electron donor P3HT mixed with an electron acceptor fullerene derivative (PCBM). In PFO/PMMA thin films, pump-probe confocal microscopy was used to examine phase separation and interfacial mixing. The authors verified that at the interface between the PFO and PMMA domains there is an efficient mixing of the two polymers, resulting in the isolation of PFO chains (Virgili et al., 2012). The same behavior was also observed in 2010 (Polli et al., 2010).

In P3HT/PCBM heterojunction, it was also verified a distinct dynamical behavior at the interface between large P3HT-rich and PCBM-rich domains. Figure 4.12(a) shows the morphology of a P3HT:PCBM blend. The whitish areas represent the PCBM-rich crystals, while the dark areas indicate the P3HT-rich domains. The $\Delta T/T$ (x, y, $\tau = 200$ fs) map reveals regions of positive signal (red) embedded in a mainly negative background (blue/black). Three different dynamics have been recorded in Figure 4.12(b): (A) PCBM-rich aggregate region, (B) interface region and (C) P3HT-rich crystal. The authors observed that the $\Delta T/T$ dynamics ($\lambda = 640$ nm) recorded at the interface presented a long-lived photoinduced absorption (PA) signal attributed to the formation of a CTS (Figure 4.12(c)). In this case, the CTS consist of partially separated, coulombically bound charge pairs, where the hole is localized on the P3HT, and the electron is on the PCBM. This behavior is necessary for the formation of separated charge carriers in bulk heterojunction solar cells (Virgili et al., 2012). A similar study of the photophysics of a crystalline phase-separated P3HT:PCBM blend with a coarsened morphology was conducted, by mapping the transient absorption signal with submicrometer space and subpicosecond time resolution (Grancini et al., 2011).

All these fluorescence spectroscopy and microscopy techniques open new possibilities towards a better understanding of the relations established between chemical structure, morphology, and photophysical properties of the conjugated polymers. The versatility and feasibility of these characterization tools allow the development of new research involving conjugated polymer chain isolation and intimate polymer intermixing, which are crucial for the optimization of organic optoelectronic devices.

4.5 INFLUENCE OF THE POLYMER STRUCTURE ON PL AND EL SPECTRA

The correlations between light emission processes and chemical structures of conjugated polymers are of crucial importance for the comprehension of the main phenomena that occur in optoelectronic devices.

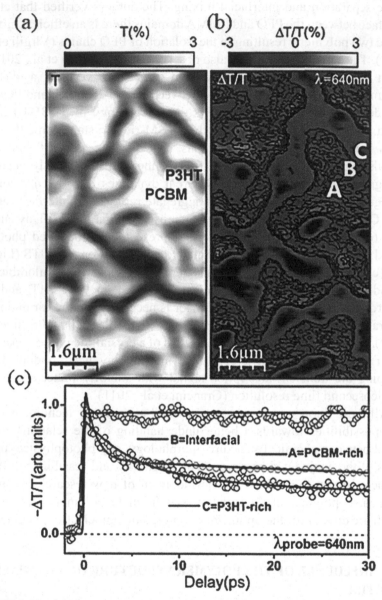

FIGURE 4.12 (a) Linear transmission image and (b) $\Delta T/T(x, y, \tau = 200\ \text{fs})$ map at $\lambda = 640$ nm for the P3HT/PCBM blend. (c) $\Delta T/T$ dynamics at $\lambda = 640$ nm for the PCBM-rich aggregate region (point A), the interfacial region (point B) and the P3HT-rich crystal (point C).

Source: Reprinted with permission from Virgili et al. (2012). Copyright 2012, Royal Society of Chemistry.

The emission color in polymeric OLEDs, for example, depends on the bandgap energy (E_{gap}) of the π–π* transition, which is determined by the polymer structure. Modifications with any specific purpose will affect E_{gap} and consequently the emitted color in the diodes (Akcelrud et al., 2003). The confinement of the conjugation into a well-defined length of the polymer chain (Chen et al., 2011) and the introduction of chromophores into the polymeric main chain (Bai et al., 2017) or side chain (Yang et al., 2014) are some of the most successful strategies developed so far to tune the optical/ electrical characteristics of these materials. In this section, some works about the influence of the chemical structure of conjugated polymers on the PL and EL spectra will be a highlight.

In a recent work, the color tuning in carbazole-functionalized homo-polymer poly[2',7'-bis(3,6-dioctylcarbazo-9-yl)-spirobifluorene] (PCzSF) through the incorporation of an additional dibenzothiophene-S,S-dioxide (3,7SO), 4,7-diphenylbenzothiadiazole (DPBT) or 4,7-dithienylbenzo-thiadiazole (DTBT) unit into its backbone was demonstrated (Bai et al., 2017). A series of polyspirobifluorene (PSF)-based red-, green-, and blue-emitting (RGB) polymers were synthesized via Suzuki polycondensation (their molecular structures are shown in Figure 4.13(a)).

(a)

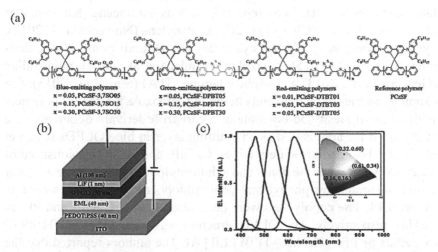

FIGURE 4.13   (a) Chemical structures of the blue-, green-, and red-emitting PSFs together with the reference polymer PCzSF. (b) Device configuration: glass-ITO | PEDOT:PSS | EML | SPPO13 | LiF | Al. (c) EL spectra at 6 V. Inset: CIE chromaticity diagram.

Source: Reprinted with permission from Bai et al. (2017). Copyright 2012, American Chemical Society.

The OLEDs were fabricated with the following configuration: glass-ITO | PEDOT:PSS | polymer | SPPO13 | LiF | Al (Figure 4.13(b)). Herein, PEDOT:PSS serves as the hole injection layer (HIL), while SPPO13 represents 9,9'-spirobi-(fluorene)-2,7-diylbis(diphenylphosphine oxide) and acts as the electron-transporting layer (ETL). The incorporated 3,7SO (DPBT or DTBT) moiety into the PCzSF may facilitate electron injection and transport. The authors also reported that the OLEDs based on PCzSF-3,7SO15, PCzSF-DPBT15, and PCzSF-DTBT03 had the best blue, green, and red device performance, revealing luminous efficiencies as high as 5.6, 21.6, and 4.4 cd A^{-1}, respectively. The EL spectra (Figure 4.13(c)) showed the signals at 460, 528, and 625 nm for PCzSF-3,7SO15, PCzSF-DPBT15 and PCzSF-DTBT03, corresponding to CIE coordinates of (0.16, 0.16), (0.32, 0.60) and (0.61, 0.34), respectively (see the inset in Figure 4.13(c)) (Bai et al., 2017). Thus, it was demonstrated the variation of the emission wavelengths for EL with the modification of the chemical structure of the polymers.

The development of donor-acceptor (D-A) type polymers, alternating electron-rich, and electron-deficient substituents along a polymer backbone, is a well-known approach to obtain efficient optoelectronics devices (Murali et al., 2012). In recent years, a variety of chromophores intercalated with fluorene has been reported, such as anthracene (Kimyonok et al., 2014), carbazole (Chen et al., 2011), phenylene (Nowacki et al., 2012), pyridine (Wang et al., 2016), cyanovinylene (Murali et al., 2013), thiophene (Sun et al., 2015), pyrazoline (Vandana et al., 2017), quinoxaline (Gedefaw et al., 2013) and triphenylamine (TPA) (Sun et al., 2015). For example, a series of random and alternating carbazole/fluorene copolymers with various dimesitylboron-containing carbazole derivative contents were synthesized for application as an emitting layer in blue OLEDs (Chen et al., 2011). Two carbazole derivatives, CzPhB, and CzPhThB consisted of a carbazolyl group as the donor and a dimesitylboron group as the acceptor group, separated by phenyl and phenylthiophene groups, as shown in Figure 4.14. The copolymers were denominated as PFCzPhB and PFCz-PhThB, respectively. The OLED structure was glass-ITO|PEDOT:PSS | PFCzPhB or PFCzPhThB | TPBi | LiF| Al. The authors reported that the electroluminescent properties of the devices were strongly dependent on the effective conjugation length and the content of the carbazole π-boron pendant. The phenyl/thiophene-linked CzPhThB unit possesses a longer effective conjugation length than the phenyl-linked CzPhB unit (see Figure

4.14). Thus, the best performance was obtained with the PFCzPhThB10 copolymer with a maximum luminance and current efficiency of 445 cd m^{-2} and 0.51 cd A^{-1}, respectively, with CIE coordinates (0.16, 0.11). On the other hand, the lower values of brightness and current efficiency were observed for the devices fabricated from the PFCzPhThB copolymers with higher CzPhThB contents, where such results were due to interruption of the π-conjugation along the polymer backbone via the incorporation of a high meta-linkage carbazole moiety content (Chen et al., 2011).

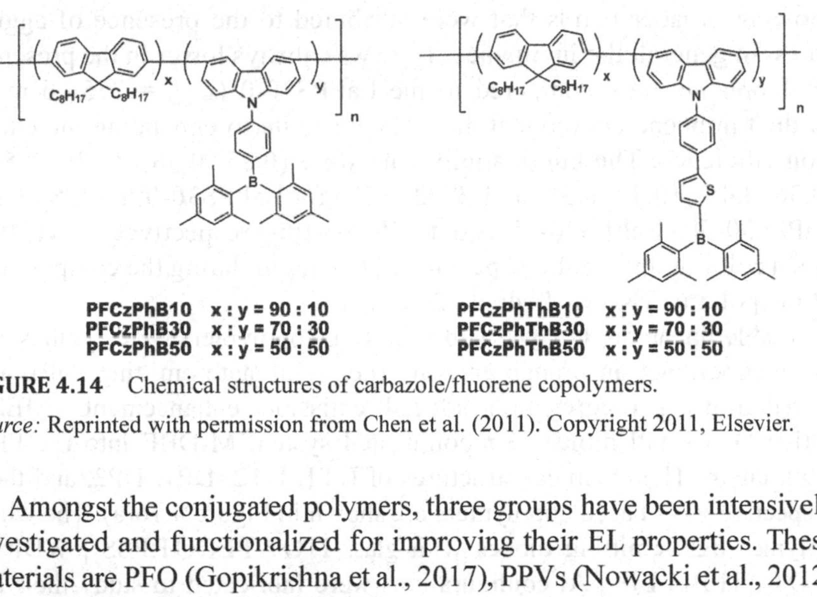

PFCzPhB10 x : y = 90 : 10	PFCzPhThB10 x : y = 90 : 10
PFCzPhB30 x : y = 70 : 30	PFCzPhThB30 x : y = 70 : 30
PFCzPhB50 x : y = 50 : 50	PFCzPhThB50 x : y = 50 : 50

FIGURE 4.14   Chemical structures of carbazole/fluorene copolymers.

Source: Reprinted with permission from Chen et al. (2011). Copyright 2011, Elsevier.

Amongst the conjugated polymers, three groups have been intensively investigated and functionalized for improving their EL properties. These materials are PFO (Gopikrishna et al., 2017), PPVs (Nowacki et al., 2012) and polycarbazoles (He et al., 2014).

PPV-type terpolymers containing fluorene, thiophene, and phenylene units were synthesized and evaluated in terms of chemical composition, PL, and EL properties. Figure 4.15(a) shows the chemical structures of the terpolymers poly[(9,9-dihexyl-9H-fluorene-2,7-diyl)-1,2-ethenediyl-1,4-phenylene-1,2-ethenediyl]x-alt-[(9,9'-dihexyl-9H-fluorene-2,7-diyl)-1,2-ethenediyl-2,5-thiophene-1,2-ethenediyl] y (LaPPS30-X) (Nowacki et al., 2012). The authors reported that the optical and electronic properties of these copolymers might be modified by changing the relative ratio of its fluorene-vinylene-phenylene units. The ratio of thiophene-vinylene/phenylene-vinylene was varied in the range 25, 50, and 75% and the molar composition of fluorene groups was maintained at 50%. These two units have different optical band gaps (E_{gap}). Thus, energy transfer by different mechanisms from the donor (fluorene-vinylene) to the acceptor (thiophene-vinylene) units may occur in the system. The digital photograph of the respective OLEDs is shown in Figure 4.15(b), and the steady-state PL spectra of the terpolymers in the solid-state and EL spectra of the glass-ITO | PEDOT:PSS | LaPPS30-X | Ca | Al diodes are depicted in Figure 4.15(c, d). The EL emissions of the terpolymers LaPPS30-X are in the same spectral range as the PL, but the emission profiles are different, showing broader bands that were attributed to the presence of aggregates. In general, the luminance (L_{max}) was always lower in the presence of thiophene units compared to the LaPPS30-0 (L_{max} = 255 cd m^{-2}), so the thiophene component did not contribute to enhancing the emission efficiency. The chromaticity data were (0.25; 0.38), (0.30; 0.55), (0.36; 0.45), (0.40; 0.56) and (0.42; 0.52) for LaPPS30-0, LaPPS30-25, LaPPS30-50, LaPPS30-75 and LaPPS30-100, respectively, revealing that a color tuning could be performed by manipulating the composition of the polymer (Nowacki et al., 2012).

Stable linear single layer white emitting copolymers were synthesized by incorporating an orange emitting (i.e., dual state emitting, DP), and a red-emitting (aggregation-induced emission enhancement (AIEE), active DT) small molecule π-conjugated system M-DBF into the PFO main chain. The chemical structures of DT1, DT2, DP1, DP2, and their respective PFO-based copolymers are shown in Figure 4.16(a). The white polymer light-emitting diodes with glass-ITO | PEDOT:PSS | WDP or WDT | TPBi | LiF | Al configurations were fabricated to study their EL properties. Figure 4.16(b, c) shows the EL characteristics of these diodes. The EL spectra exhibited a slight difference in the blue region as compared with PL spectra (data not shown). The copolymers showed dual emission signals at the blue region from the PFO (432 and 462 nm) and the orange region from the M-DBFs (554–574 nm for WDP or 578–604 nm for WDT), respectively. In general, the devices fabricated using WDP-1 or WDT-1 copolymers were found to emit white emission with CIE coordinates of (0.31, 0.33) and (0.35, 0.34) that are very close to standard CIE coordinates (0.33, 0.33) with maximum luminous efficiencies of 7.82 and 4.57 cd A^{-1}, with highest brightness values of 9753 and 7436 cd m^{-2} (Gopikrishna et al., 2017).

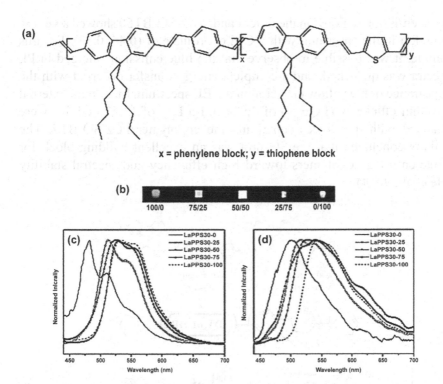

FIGURE 4.15 (a) Chemical structures of the LaPPS30 series, (b) digital photograph of the OLEDs with the configuration: glass-ITO | PEDOT:PSS | LaPPS30-X | Ca | Al, (c) steady-state PL spectra of the LaPPS30 series in the film form, (d) EL of the copolymers of the LaPPS30 series.

Source: Reprinted with permission from Nowacki et al. (2012). Copyright 2012, Elsevier.

White-emitting copolymers (PCz-SO-BT)s, based on 2,7-carbazole (Cz), dibenzothiophene-S,S-dioxide (SO) (blue emitter) and benzo-thiadiazole (BT) (yellow emitter) units, as shown in Figure 4.17(a), have also been investigated (He et al., 2014). OLEDs based on the resulting polymers were fabricated using the structure glass-ITO | PEDOT:PSS | polymer | CsF | Al. The PL and EL spectra are shown in Figure 4.17(b, c). The EL spectra of the diodes presented stronger emission at the longer wavelength band, broader half-bandwidth, and a little redshift in comparison to the PL of the polymeric films. The CIE coordinate of the polymers PCz-BT4, PCz-SO-BT7, and PCz-SO-BT10 were (0.32, 0.38), (0.32, 0.41), and (0.30, 0.45), respectively, which all correspond to the white emission region. On the other hand, PCz-SO-BT50 showed a yellow emission in the spectrum, with a CIE coordinate of (0.44, 0.53). For this sample, it was possible to observe that the blue emission emerged in PL spectra was quenched, and a complete energy transfer occurred with the appearance of the yellow emission in the EL spectrum. Maximum external quantum efficiency (EQE_{max}) of 3.9% and a L_{max} of 13 526 cd m^{-2} were obtained with the diodes containing the copolymer PCz-SO-BT7. The authors concluded that the SO unit was an excellent building block for white-emitting copolymers toward high efficiency and spectral stability (He et al., 2014).

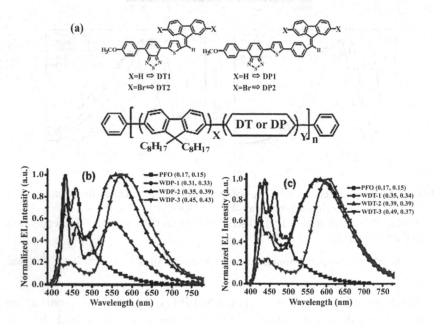

FIGURE 4.16 (a) Chemical structures of DT1, DT2, DP1, DP2, and their respective PFO-based copolymers. EL spectra of (b) WDP and (c) WDT-based OLEDs.

Source: Reprinted with permission from Gopikrishna et al. (2017). Copyright 2017, American Chemical Society.

In another example, a methodology for modifying PFO structures using icosahedral carboranes ($C_2B_{10}H_{12}$) was proposed (Davis et al., 2012). The authors synthesized *o*-carborane-containing poly(fluorene) (PF-*o*-carb) and fluorene-thiophene copolymer incorporating *o*-carborane

(PFT-*o*-carb). The chemical structures are illustrated in Figure 4.18(a). OLEDs were fabricated from each of the polymers with the general device architecture glass-ITO | PEDOT:PSS | emissive polymer | Ba | Al. PL and EL spectra of PF, PFT, and their carborane-containing derivatives are shown in Figure 18(b, c). The EL spectra of the diodes are comparable to the PL spectra of the polymer films, showing large redshifting of PF-*o*-carb and PFT-*o*-carb emission compared to the polymers without carborane (PF and PFT). According to the authors, the carborane-containing derivatives ($E_{gap(PF\text{-}o\text{-}carb)}$ = 3.23 eV and $E_{gap(PFT\text{-}o\text{-}carb)}$ = 2.81 eV) presented increased band gaps (E_{gap}) compared to their PF and PFT analogs ($E_{gap(PF)}$ = 2.96 eV and $E_{gap(PFT)}$ = 2.58 eV), with a significant influence on the optical properties of these materials. It was also reported the occurrence of the cages' effect in *o*-carborane-containing polymers. The peak emission wavelengths for EL (λ_{EL}) were 432, 570, 517, and 620 nm for PF, PF-*o*-carb, PFT, and PFT-*o*-carb, respectively. The inset in Figure 4.18(c) shows digital photographs of operating OLEDs (Davis et al., 2012).

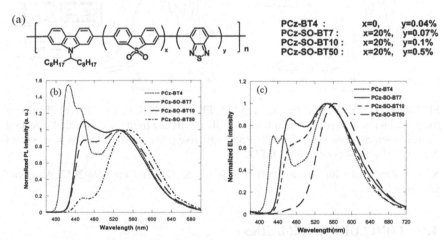

FIGURE 4.17   (a) Chemical structures of the copolymers PCz-BT and PCz-SO-BT, (b) PL spectra of the polymers in the film, (c) EL spectra of the polymers at 12 mA m^{-2}.

Source: Reprinted with permission from He et al. (2014). Copyright 2012, Royal Society of Chemistry.

To summarize, Section 4.4 showed that there are many effective strategies to tune the PL and EL of conjugated polymers. The discussed examples illustrate a variety of chemical structure modifications that can be used for the design of SemPolys with light emissions covering the entire visible wavelength region. Hence, these materials might fulfill the need for color tuning in applications involving PL and EL phenomena.

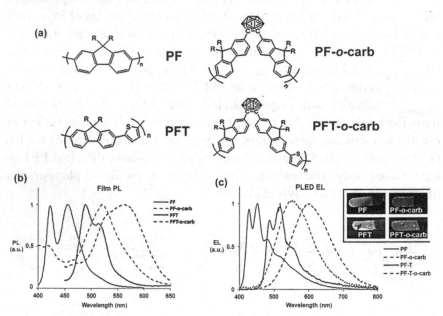

FIGURE 4.18 (a) Chemical structures of PF, PF-o-carb, PFT, PFT-o-carb, (b) PL spectra for poly(fluorene) derivatives thin films, (c) EL spectra are from OLEDs with the general device structure: glass-ITO | PEDOT: PSS | emissive polymer | Ba | Al. The inset shows digital photographs of operating OLEDs.

Source: Reprinted with permission from Davis et al. (2012). Copyright 2012, American Chemical Society.

4.6 CONCLUDING REMARKS

The comprehension of PL and EL processes of polymers is of great interest since these phenomena can be explored in the development of materials for a variety of applications. The most divergent point between the PL and EL phenomena is related to the formation of the excited states: by photo-excitation (in PL), or by charge carrier recombination (In EL). In the EL process, the electron-hole capture is generally a spin-independent process, whereas, in the PL process, the occurrence of singlet or triplet excited states follows the spin multiplicity rule. Such differences are expected to affect quantum emission efficiency. Investigations of the correlation between the two phenomena facilitate the design of novel materials with controlled properties, to provide emission in a specific spectral region, with high quantum efficiency. In several cases, the PL and EL spectra of a semiconducting polymer are similar. Thus, the study of photoluminescent behavior can give a prediction of what will happen in the electroluminescent emissive process. For example, if the material has a low quantum efficiency in PL, EL efficiency will also not be high. However, it is not possible to guarantee that a material with high PL efficiency will also have a high EL efficiency.

Furthermore, in the solid-state, the PL and EL spectra of a polymer, or of polymer blends, can present different features resulting from the existence of non-interacting isolated chain segments and aggregated segments, showing varying degrees of intrachain and interchain interactions. Thus, relevant information can be driven by the analysis of the fluorescent behavior of polymers, including the assessment of the organization/structure of polymeric films. It has also been discussed how to tune the PL and EL of conjugated polymers by modifications of their chemical structure, for the design of SemPolys with emissions covering the entire visible wavelength region and hence, successfully fulfill the need for color tuning and preparation of efficient light-emitting diodes.

KEYWORDS

- **confocal scanning microscopy**
- **electroluminescence**
- **fluorescence correlation spectroscopy**
- **photoluminescence**
- **pump-probe confocal microscopy**
- **single-molecule fluorescence spectroscopy**
- **transient absorption spectroscopy**
- **wide-field fluorescence microscopy**

REFERENCES

Adachi, T., Vogelsang, J., & Lupton, J. M., (2014). Unraveling the electronic heterogeneity of charge traps in conjugated polymers by single-molecule spectroscopy. *J. Phys. Chem. Lett., 5*, 573–577.

Akcelrud, L., (2003). Electroluminescent polymers. *Prog. Polym. Sci., 28*, 875–962.

Bai, K., Wang, S., Zhao, L., Ding, J., & Wang, L., (2017). Efficient blue, green, and red electroluminescence from carbazole-functionalized poly(spirobifluorene)s. *Macromolecules, 50*, 6945–6953.

Barbara, P. F., Gesquiere, A. J., Park, S. J., & Lee, Y. J., (2005). Single-molecule spectroscopy of conjugated polymers. *Acc. Chem. Res., 83*, 602–610.

Becker, K., & Lupton, J. M., (2005). Dual species emission from single polyfluorene molecules: Signatures of stress-induced planarization of single polymer chains. *J. Am. Chem. Soc., 127*, 7306–7307.

Birks, J. B., (1970). *Photophysics of Aromatic Molecules*. John Wiley & Sons Ltd: New York.

Bolinger, J. C., Traub, M. C., Brazard, J., Adachi, T., Barbara, P. F., & Bout, D. A. V., (2012). Conformation and energy transfer in single conjugated polymers. *Acc. Chem. Res., 45*, 1992–2001.

Botiz, I., Astilean, S., &Stingelin, N., (2016). Altering the emission properties of conjugated polymers. *Polym. Int., 65*, 157–163.

Buckley, A. R., Rahn, M. D., Hill, J., Cabanillas-Gonzalez, J., Fox, A. M., & Bradley, D. D. C., (2011). Energy transfer dynamics in polyfluorene-based polymer blends. *Chem. Phys. Lett., 339*, 331–336.

Burroughes, J. H., Bradley, D. D. C., Brown, A. R., Marks, R. N., Mackay, K., Friend, R. H., Burns, P. L., & Holmes, A. B., (1990). Light-emitting diodes based on conjugated polymers. *Nature, 347*, 539–541.

Cabanillas-Gonzalez, J., Grancini, G., &Lanzani, G., (2011). Pump-probe spectroscopy in organic semiconductors: Monitoring fundamental processes of relevance in optoelectronics. *Adv. Mater., 23*, 5468–5485.

Chappell, J., Lidzey, D. G., Jukes, P. C., Higgins, A. M., Thompson, R. L., O'Connor, S., Grizzi, I., et al., (2003). Correlating structure with fluorescence emission in phase-separated conjugated-polymer blends, *Nat. Mater., 2*, 616–621.

Chen, R., Wu, R., Zhang, G., Gao, Y., Xiao, L., & Jia, S., (2014). Electron transfer-based single molecule fluorescence as a probe for nano-environment dynamics. *Sensors, 14*, 2449–2467.

Chen, S. A., Jen, T. H., & Lu, H. H., (2010). A review on the emitting species in conjugated polymers for photo-and electro-luminescence. *J. Chin. Chem. Soc., 57*, 439–458.

Chen, Y. H., Lin, Y. Y., Chen, Y. C., Lin, J. T., Lee, R. H., Kuo, W. J., &Jeng, R. J., (2011). Carbazole/fluorene copolymers with dimesitylboron pendants for blue light-emitting diodes. *Polymer, 52*, 976–986.

Chua, L. L., Zaumseil, J., Chang, J. F., Ou, E. C. W., Ho, P. K. H., Sirringhaus, H., & Friend, R. H., (2005). General observation of n-type field-effect behavior in organic semiconductors. *Nature, 434*, 194–199.

Como, E. D., Becker, K., & Lupton, J. M., (2008). Single molecule spectroscopy of polyfluorenes. *Adv. Polym. Sci., 212*, 293–318.

Como, E. D., Borys, N. J., Strohriegl, P., Walter, M. J., & Lupton, J. M., (2011). Formation of a defect-free π-electron system in single β-phase polyfluorene chains. *J. Am. Chem. Soc., 133*, 3690–3692.

Davis, A. R., Peterson, J. J., & Carter, K. R., (2012). Effect of o-carborane on the optoelectronic and device-level properties of poly(fluorene)s. *Acs. Macro Lett.*, 469–472.

Dimitrov, S. D., Bakulin, A. A., Nielsen, C. B., Schroeder, B. C., Du, J., Bronstein, H., McCulloch, I., et al., (2012). On the energetic dependence of charge separation in low-bandgap polymer/fullerene blends. *J. Am. Chem. Soc., 134*, 18189–18192.

Flier, B. M. I., Baier, M., Huber, J., Müllen, K., Mecking, S., Zumbusch, A., &Wöll, D., (2011). Single molecule fluorescence microscopy investigations on the heterogeneity of translational diffusion in thin polymer films. *Phys. Chem. Chem. Phys., 13*, 1770–1775.

Friend, R. H., Gymer, R. W., Holmes, A. B., Burroughes, J. H., Marks, R. N., Taliani, C., Bradley, D. D. C., et al., (1999). Electroluminescence in conjugated polymers. *Nature, 397*, 121–128.

Gedefaw, D., Ma, Z., Mulugeta, E., Zhao, Y., Zhang, F., Andersson, M. R., &Mammo, W., (2013). An alternating copolymer of fluorene donor and quinoxaline acceptor versus a terpolymer consisting of fluorene, quinoxaline, and benzothiadiazole building units: Synthesis and characterization. *Polym. Bull., 73*, 1167–1183.

Ghosh, A., Jana, B., Chakraborty, S., Maiti, S., Jana, B., Ghosh, H. N., & Patra, A., (2017). Exciton dynamics and formation mechanism of MEH-PPV polymer-based nanostructures. *J. Phys. Chem. C, 212*, 21062–21072.

Giro, G., Cocchi, M., Kalinowski, J., Di Marco, P., &Fattori, V., (2000). Multi-component emission from organic light emitting diodes based on polymer dispersion of an aromatic diamine and an oxadiazole derivative. *Chem. Phys. Lett., 318*, 137–141.

Gopikrishna, P., Das, D., Adil, L. R., &Iyer, P. K., (2017). Saturated and stable white electroluminescence from linear single polymer systems based on polyfluorene and mono-substituted dibenzofulvene derivatives. *J. Phys. Chem. C, 121*, 18137–18143.

Grancini, G., Polli, D., Fazzi, D., Cabanillas-Gonzalez, J., Cerullo, G., &Lanzani, G., (2011). Transient absorption imaging of P3HT:PCBM photovoltaic blend: Evidence for interfacial charge transfer state. *J. Phys. Chem. Lett., 2*, 1099–1105.

Grumstrup, E. M., Gabriel, M. M., Cating, E. E. M., Goethem, E. M. V., &Papanikolas, J. M., (2015). Pump-probe microscopy: Visualization and spectroscopy of ultrafast dynamics at the nanoscale. *Chem. Phys., 458*, 30–40.

Haustein, E., &Schwille, P., (2004). Single-molecule spectroscopic methods. *Curr. Opin. Struct. Biol., 14*, 531–540.

Haustein, E., &Schwille, P., (2007). Fluorescence correlation spectroscopy: Novel variations of an established technique. *Annu. Rev. Biophys. Biomol. Struct., 36*, 151–169.

Hayer, A., Khan, A. L. T., Friend, R. H., & Köhler, A., (2005). Morphology dependence of the triplet excited state formation and absorption in polyfluorene. *Phys. Rev. B, 71*, 241302.

He, R., Xu, J., Xue, Y., Chen, D., Ying, L., Yang, W., & Cao, Y., (2014). Improving the efficiency and spectral stability of white-emitting polycarbazoles by introducing a dibenzothiophene-S,S-dioxide unit into the backbone. *J. Mater. Chem. C, 2*, 7881–7890.

Honmou, Y., Hirata, S., Komiyama, H., Hiyoshi, J., Kawauchi, S., Iyoda, T., &Vacha, M., (2014). Single-molecule electroluminescence and photoluminescence of polyfluorene unveils the photophysics behind the green emission band. *Nature Commun., 5*, 1–8.

Hu, D., Yu, J., & Barbara, P. F., (1999). Single-molecule spectroscopy of the conjugated polymer MEH-PPV. *J. Am. Chem. Soc., 121*, 6936–6937.

Hu, D., Yu, J., Wong, K., Bagchi, B., Rossky, P. J., & Barbara, P. F., (2000). Collapse of stiff conjugated polymers with chemical defects into ordered, cylindrical conformations. *Nature, 405*, 1030–1033.

Huser, T., & Yan, M., (2001). Solvent-related conformational changes and aggregation of conjugated polymers studied by single-molecule fluorescence spectroscopy. *J. Photochem. Photobiol. A, 144*, 43–51.

Huser, T., Yan, M., & Rothberg, L. J., (2000). Single chain spectroscopy of conformational dependence of conjugated polymerphotophysics. *Proc. Natl. Acad. Sci. U.S.A., 97*, 11187–11191.

Jokinen, K., Bykov, A. V., Sliz, R., Remes, K., Fabritius, T., &Myllyla, R., (2015). Light emission color conversion of polyfluorene-blend OLEDs induced by thermal annealing. *IEEE Trans. Electron Devices, 62*, 2238–2243.

Jokinen, K., Bykov, A., Sliz, R., Remes, K., Fabritius, T., &Myllyla, R., (2015). Luminescence and spectrum variations caused by thermal annealing in undoped and doped polyfluorene OLEDs. *Solid-State Electron, 103*, 184–189.

Kaake, L. G., Jasieniak, J. J., Bakus, R. C., Welch, G. C., Moses, D., Bazan, G. C., &Heeger, A. J., (2012). Photo induced charge generation in a molecular bulk heterojunction material. *J. Am. Chem. Soc., 134*, 19828–19838.

Kalinowski, J., (2008). Bimolecular excited species in optical emission from organic electroluminescent devices. *J. Non-Cryst. Solids, 354*, 4170–4175.

Kalinowski, J., Giro, G., Cocchi, M., Fattori, V., & Di Marco, P., (2000). Unusual disparity in electroluminescence and photoluminescence spectra of vacuum-evaporated films of 1,1-bi-di-4-tolylamino-phenyl-cyclohexane. *Appl. Phys. Lett., 76*, 2352.

Kee, T. W., (2014). Femtosecond pump-push-probe and pump-dump-probe spectroscopy of conjugated polymers: New insight and opportunities. *J. Phys. Chem. Lett., 5*, 3231–3240.

Khan, A. L. T., Sreearunothai, P., Herz, L. M., Banach, M. J., & Köhler, A., (2004). Morphology-dependent energy transfer within polyfluorene thin films. *Phys. Rev. B., 69*, 085201.

Kimyonok, A., Tekin, E., Haykır, G., &Turksoy, F., (2014). Synthesis, photophysical and electroluminescence properties of anthracene-based green-emitting conjugated polymers. *J. Lumin., 146*, 186–192.

Kondo, T., Chen, W. J., & Schlau-Cohen, G. S., (2017). Single-molecule fluorescence spectroscopy of photosynthetic systems. *Chem. Rev., 117*, 860–898.

Koynov, K., & Butt, H. J., (2012). Fluorescence correlation spectroscopy in colloid and interface science.*Curr. Opin. Colloid Interface Sci., 17*, 377–387.

Lakowicz, J. R., (1999). *Principles of Fluorescence Spectroscopy* (2ndedn.). Plenum Publishing Corporation.

List, E. J. W., Guentner, R., Freitas, P. S. D., &Scherf, U., (2002). The effect of keto defect sites on the emission properties of polyfluorene-type materials. *Adv. Mater., 14*, 374–378.

Lupton, J. M., (2010). Single-molecule spectroscopy for plastic electronics: Materials analysis from the bottom-up. *Adv. Mater., 22*, 1689–1721.

Moerner, W. E., & Fromm, D. P., (2003). Methods of single-molecule fluorescence spectroscopy and microscopy. *Rev. Sci. Instrum., 74*, 3597–3619.

Murali, M. G., Dalimba, U., Yadav, V., & Srivastava, R., (2013). New thiophene-based donor-acceptor conjugated polymers carrying fluorene or cyanovinylene units: Synthesis, characterization, and electroluminescent properties. *Polym. Eng. Sci., 53*, 1161–1170.

Murali, M. G., Naveen, P., Udayakumar, D., Yadav, V., & Srivastava, R., (2012). Synthesis and characterization of thiophene and a fluorene-based donor-acceptor conjugated polymer containing 1,3,4-oxadiazole units for light-emitting diodes. *Tetrahedron Lett., 53*, 157–161.

Nowacki, B., Grova, I. R., Domingues, R. A., Faria, G. C., Atvars, T. D. Z., &Akcelruda, L., (2012). Photo-and electroluminescence in a series of PPV type terpolymers containing fluorene, thiophene and phenylene units. *J. Photochem. Photobiol. A., 237*, 71–79.

Ohkita, H., & Ito, S., (2011). Transient absorption spectroscopy of polymer-based thin-film solar cells. *Polymer, 52*, 4397–4417.

Ohkita, H., Cook, S., Ford, T. A., Greenham, N. C., &Durrant, J. R., (2006). Monomolecular triplet decay dynamics in fluorene-based conjugated polymer films studied by transient absorption spectroscopy. *J. Photochem. Photobiol. A, 182*, 225–230.

Ohkita, H., Tamai, Y., Benten, H., & Ito, S., (2016). Transient absorption spectroscopy for polymer solar cells. *IEEE J. Sel. Topics Quantum Electron, 22*, 4100612.

Orrit, M., Ha, T., &Sandoghdar, V., (2014). Single-molecule optical spectroscopy. *Chem. Soc. Rev., 43*, 973–976.

Ostroverkhova, O., (2016). Organic optoelectronic materials: Mechanisms and applications. *Chem. Rev., 116*, 13279–13412.

Papadakis, C. M., Košovan, P., Richtering, W., &Wöll, D., (2014). Polymers in focus: Fluorescence correlation spectroscopy. *Colloid Polym. Sci., 292*, 2399–2411.

Pereira, J., Farinhas, J., &Morgado, J., (2016). Suppressing the energy transfer in polymer blends films upon addition of a co-solvent, *Mater. Lett., 15*, 248–251.

Polli, D., Grancini, G., Clark, J., Celebrano, M., Virgili, T., Cerullo, G., &Lanzani, G., (2010). Nanoscale imaging of the interface dynamics in polymer blends by femtosecond pump-probe confocal microscopy. *Adv. Mater., 22*, 3048–3051.

Quites, F. J., Faria, G. C., Germino, J. C., &Atvars, T. D. Z., (2014). Tuning emission colors from blue to green in polymericlight-emitting diodes fabricated using polyfluorene blends. *J. Phys. Chem. A, 118*, 10380–10390.

Reid, O. G., Rayermann, G. E., Coffey, D. C., & Ginger, D. S., (2010). Imaging local trap formation in conjugated polymer solar cells: A comparison of time-resolved electrostatic force microscopy and scanning kelvin probe imaging. *J. Phys. Chem. C, 114*, 20672–20677.

Sartori, S. S., Feyter, S. D., Hofkens, J., Auweraer, M. V. D., Schryver, F. D., Brunner, K., &Hofstraat, J. W., (2003). Host matrix dependence on the photophysical properties of individual conjugated polymer chains. *Macromolecules, 36*, 500–507.

Stangl, T., Wilhelm, P., Remmerssen, K., Höger, S., Vogelsang, J., & Lupton, J. M., (2015). Mesoscopic quantum emitters from deterministic aggregates of conjugated polymers. *Proc. Natl. Acad. Sci. U.S.A., 112*, E5560–E5566.

Steiner, F., Lupton, J. M., & Vogelsang, J., (2017). Role of triplet-state shelving in organic photovoltaics: Single-chain aggregates of poly(3-hexylthiophene) versus mesoscopic multichain aggregates. *J. Am. Chem. Soc., 139*, 9787–9790.

Steiner, F., Vogelsang, J., & Lupton, J. M., (2014). Singlet-triplet annihilation limits exciton yield in poly(3-Hexylthiophene). *Phys. Rev. Lett., 112*, 137402.

Sun, M. M., Wang, W., Liang, L. Y., Yan, S. H., Zhou, M. L., & Ling, Q. D., (2015). Substituent effects on direct arylation polycondensation and optical properties of alternating fluorene-thiophene copolymers. *Chin. J. Polym. Sci., 33*, 783–791.

Sun, N., Feng, F., Wang, D., Zhou, Z., Guan, Y., Dang, G., Zhou, H., Chen, C., & Zhao, X., (2015). Novel polyamides with fluorene-based triphenylamine: Electrofluorescence and electrochromic properties. *RSC Advances, 5*, 88181–88190.

Thiessen, A., Vogelsang, J., Adachi, T., Steiner, F., Bout, D. V., & Lupton, J. M., (2013). Unraveling the chromophoric disorder of poly(3-hexylthiophene). *Proc. Natl. Acad. Sci. U.S.A., 110*, E3550-E3556.

Vacha, M., &Habuchi, S., (2010). Conformation, and physics of polymer chains: A single-molecule perspective. *Npg Asia Mater., 2*, 134–142.

Vandana, T., Karuppusamy, A., & Kannan, P., (2017). Polythiophenylpyrazoline containing fluorene and benzothiadiazole moieties as blue and white light emitting materials. *Polymer, 124*, 88–94.

Virgili, T., Grancini, G., Molotokaite, E., Suarez-Lopez, I., Rajendran, S. K., Liscio, A., Palermo, V., Lanzani, G., et al., (2012). Confocal ultrafast pump-probe spectroscopy: A new technique to explore nanoscale composites. *Nanoscale, 4*, 2219–2226.

Virgili, T., Lidzey, D. G., Grell, M., Walker, S., Asimakis, A., & Bradley, D. D. C., (2001). Completely polarized photoluminescence emission from a micro cavity containing an aligned conjugated polymer. *Chem. Phys. Lett., 341*, 219–224.

Vogelsang, J., & Lupton, J. M., (2012). Solvent vapor annealing of single conjugated polymer chains: Building organic optoelectronic materials from the bottom up. *J. Phys. Chem. Lett., 3*, 1503–1513.

Wang, J., Bao, X., Ding, D., Qiu, M., Du, Z., Wang, J., Liu, J., Sun, M., & Yang, R., (2016). A fluorine-induced high-performance narrow bandgap polymer based on thiadiazolo[3, 4-c] pyridine for photovoltaic applications. *J. Mater. Chem. A., 4*, 11729–11737.

White, J. D., Hsu, J. H., Wang, C. F., Chen, Y. C., Hsiang, J. C., Su, S. C., Sun, W. Y., &Fann, W. S., (2002). Single-molecule fluorescence spectroscopy. *J. Chin. Chem. Soc., 49*, 669–676.

Wilson, J. S., Dhoot, A. S., Seeley, A. J. A. B., Khan, M. S., Köhler, A., & Friend, R. H., (2001). Spin-dependent exciton formation in π-conjugated compounds, *Nature, 413*, 828–831.

Wöll, D., (2014). Fluorescence correlation spectroscopy in polymer science. *RSC Advanced, 4*, 2447–2465.

Wöll, D., Braeken, E., Deres, A., Schryver, F. C. D., Uji-i, H., &Hofkens, J., (2009). Polymers and single-molecule fluorescence spectroscopy, what can we learn? *Chem. Soc. Rev., 38*, 313–328.

Wöll, D., Uji-i, H., Schnitzler, T., Hotta, J. I., Dedecker, P., Herrmann, A., Schryver, F. C. D., et al., (2008). Radical polymerization tracked by single molecule spectroscopy. *Angew. Chem. Int. Ed., 47*, 783–787.

Wu, E. C., Stubbs, R. E., Peteanu, L. A., Jemison, R., McCullough, R. D., &Wildeman, J., (2017). Detection of ultralow concentrations of non-emissive conjugated polymer aggregates via fluorescence correlation spectroscopy. *J. Phys. Chem. B., 121*, 5413–5421.

Yamamoto, S., Ohkita, H., Benten, H., & Ito, S., (2012). Role of interfacial charge transfer state in charge generation and recombination in low-band gap polymer solar cell. *J. Phys. Chem. C, 116,* 14804–14810.

Yang, J., Park, H., & Kaufman, L. J., (2017). Highly anisotropic conjugated polymer aggregates: Preparation and quantification of physical and optical anisotropy. *J. Phys. Chem. C, 121,* 13854–13862.

Yang, Y., Yu, L., Xue, Y., Zou, Q., Zhang, B., Ying, L., Yang, W., Peng, J., & Cao, Y., (2014). Improved electroluminescence efficiency of polyfluorenes by simultaneously incorporating dibenzothiophene-S,S-dioxide unit in main chain and oxadiazole moiety in side chain. *Polymer, 55,* 1698–1706.

Yanga, S. Y., Zhang, X. L., Lou, Z. D., & Hou, Y. B., (2007). Charge tunneling and cross recombination at organic heterojunction under electric fields. *Eur. Phys. J. B, 59,* 151.

Yoo, H., Furumaki, S., Yang, J., Lee, J. E., Chung, H., Oba, T., Kobayashi, H., et al., (2012). Excitonic coupling in linear and trefoil trimer perylenediimide molecules probed by single-molecule spectroscopy. *J. Phys. Chem. B, 116,* 12878–12886.

Zakya, H. K., (2005). *Organic Electroluminescence.* Taylor & Francis: New York.

CHAPTER 5

Photophysical Properties of Polythiophenes

F. S. FREITAS[1] and R. F. C. NEVES[2]

[1]The Federal Institute of Education, Science and Technology of the South of Minas Gerais, Pouso Alegre Campus, Maria da Conceição Santos Avenue 900, 37560-260, Pouso Alegre, MG, Brazil

[2]The Federal Institute of Education, Science and Technology of the South of Minas Gerais, Poços de Caldas Campus, Dirce Pereira Rosa Avenue 300, 37713-100, Poços de Caldas, MG, Brazil

5.1 INTRODUCTION

Polythiophenes (PT) have been exploited over the past several years due to their structural versatility and optical properties upon functionalization to the thiophene main chain (Jen et al., 1986; Sato et al., 1986). Its structure of C atoms may be compared as a modification of *trans* polyacetylene with a more complicated organization due to the bond with sulfur atom forming a ring, as shown in Figure 5.1. This modification becomes PT stable under atmospheric and thermal exposure, beyond the possibility of tuning crystallinity and solubility through functionalization (Springborg, 1992).

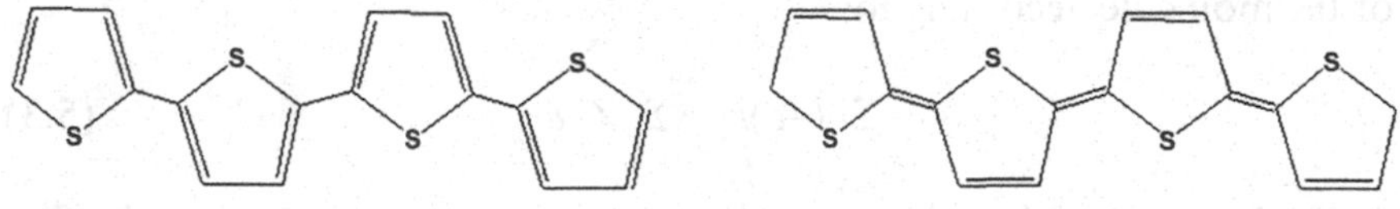

FIGURE 5.1 Representation of aromatic and quinoid structures for polythiophene, respectively.

5.2   MECHANISMS OF EXCITED STATES

PTs are conjugated polymers that can be used as molecular electronic materials in various applications principally in solar energy conversion, due to its great electrical and optical properties (Wang et al., 2014). It is remarkable how PT has reached a prominent position due to its relatively high conductivity, environmental stability of its neutral state, and its structural versatility. Their overlapping p-orbitals generate a set of delocalized π-electrons along their carbon backbone, which results in important and useful optical and electronic properties.

The achievement of a complete understanding of the relationship between the chemical structure and the electronics of the polymer is one of the most important aims in the field of electrically conducting polymers. It is necessary to analyze the interaction between the electromagnetic field and the molecular target to comprehend how optical transition energy relates to the chemical structure of the conjugated backbone and the molecular packing of these polymers.

The dipole moment of a set of point charges can be defined classically by:

$$\mu = \Sigma_i \, q_i \, r_i \tag{5.1}$$

where, r_i is the vector position of the charge q_i. If an electric field $E = \varepsilon_z k$ is applied, the charge q_i will suffer the force $F = q_i E$; for a system of several charges, the potential is:

$$V = -E \cdot \mu \tag{5.2}$$

Considering the potential as a perturbation in the molecule, it's possible to find the quantum expression of the molecular dipole moment. Then, the quantum operator can be defined as $\hat{H}' = -E \cdot \hat{\mu}$, where, $\hat{\mu}$ is the electric dipole moment operator, which can be extended for nuclei and electrons of the molecule according to:

$$\hat{\mu} = \Sigma i \, (-e) r_i + \Sigma_n Z_n \, e r_n \tag{5.3}$$

In this sense, when the system is submitted to a perturbation, the first order correction of the energy $\left\langle \psi^{(0)} | \hat{H}' | \psi^{(0)} \right\rangle$ becomes:

$$E^{(1)} = -E \cdot \left\langle \psi_{el} | \hat{\mu} | \psi_{el} \right\rangle \tag{5.4}$$

where, ψ_{el} is the electronic wave function in the absence of the field. Furthermore, the molecular quantum dipole moment μ is defined by:

$$\mu = \left\langle \psi_{el} | \hat{\mu} | \psi_{el} \right\rangle \tag{5.5}$$

Since that ψ_{el} and $\hat{\mu}$ depend on the nuclear configuration, dipole moment is a function of the nuclear configuration and it is called dipole moment function.

An electromagnetic wave (or light) is composed of mutually perpendicular oscillating electric and magnetic fields. If a monochromatic and polarized (in the xy-plane) radiation is propagating along the z-axis, the electric and magnetic fields, E, and B, can be written by:

$$E = \mathcal{E}_x (t) i = \mathcal{E}_{0x} \cos \cos \left(2\pi v t - \frac{2\pi z}{\lambda} \right) i \tag{5.6}$$

and

$$B = B_y (t) j = B_{0y} \cos \left(2\pi v t - \frac{2\pi z}{\lambda} \right) j \tag{5.7}$$

where, $\mathcal{E}_{0x}$ and B_{0y} are the maximum intensities of the electric and magnetic fields respectively, λ is the wavelength, v is the frequency so that $\lambda v = c$, where, $c = 299{,}792{,}458$ m/sec is the speed of light in vacuum.

When studying a system such as a molecule exposed to electromagnetic radiation, the quantum field theory should be the treatment of this interaction that considers both the molecule and the radiation in the quantum mechanics (QM) approach. Nevertheless, it is possible to describe this interaction by treating just the molecule quantum-mechanically. It is possible to define the perturbation part of the Hamiltonian by:

$$\hat{H}'(t) = -\mathcal{E}_{0x} \sum_i q_i x_i \cos \left(\omega t - 2\pi \frac{z_i}{\lambda} \right) \tag{5.8}$$

The PT families absorb light in the ultraviolet and visible region of the spectrum. The ultraviolet and visible range of energy is related to wavelengths much larger than the molecule size. Since the electron, movement is restricted to the molecule region, the spatial change of the light's electric field may not be considered, then:

$$\hat{H}'(t) = -\mathcal{E}_{0x} cos\omega t \sum_i q_i x_i \tag{5.9}$$

and it is possible to rewrite the previous equation by using the relation $\mu = \Sigma_i q_i r_i$, which results in:

$$\hat{H}'(t) = -\hat{\mu}_x \mathcal{E}_{0x} \frac{1}{2}\left(e^{i\omega t} + e^{-i\omega t}\right) \tag{5.10}$$

According to the time-dependent perturbation theory (Cohen-Tannoudji et al., 1992), considering a perturbation $\hat{H}'(t)$ acting from $t = 0$ up to $t = t_1$, the state function of the system can be written for a specific time t by:

$$\psi(\alpha,t) = \sum_j b_j(t_1) e^{-\frac{iE_j^{(0)}t}{\hbar}} \psi_j^{(0)}(\alpha), t \geq t_1 \tag{5.11}$$

where, $E_j^{(0)}$ and $\psi_j^{(0)}(\alpha)$ are the energy and the wave function of the stationary unperturbed states and α represents the 3n spatial coordinates and the n spin coordinates of the n-particle system. The expansion's coefficients $b_j(t)$ are determined by:

$$b_j(t_1) = b_k(0) - \frac{i}{\hbar}\int_0^{t_1} e^{i\omega_{jk}t}\psi_j^{(0)}|\hat{H}'(t)|\psi_k^{(0)} dt \tag{5.12}$$

and

$$b_k(0) = \delta_{jk} ; \omega_{jk} = \frac{E_j^{(0)} - E_k^{(0)}}{\hbar} \tag{5.13}$$

Then, the perturbation induces a transition from the k state to the j state, and the probability of this transition occurring is $|b_j(t_1)|^2$.

By using the Hamiltonian of the perturbation (Eq. (5.10)), it is possible to find:

$$b_j(t_1) = \delta_{jk} + \frac{i\mathcal{E}_{0x}}{2\hbar}\left\langle\psi_j^{(0)}|\hat{\mu}_x|\psi_k^{(0)}\right\rangle\left[\frac{e^{i(\omega_{jk}+\omega)t_1}-1}{\omega_{jk}+\omega} + \frac{e^{i(\omega_{jk}-\omega)t_1}-1}{\omega_{jk}-\omega}\right] \tag{5.14}$$

According to Eq. (5.13) and that $\omega = 2\pi v$, considering a transition from the state k to the state j which has a reasonable amplitude when, $\omega_{jk} = \omega$ this assumption leads to:

$$E_j^{(0)} - E_k^{(0)} = h\nu \qquad (5.15)$$

Therefore, according to Eq. (5.15), a system is excited from the state k to state j when irradiated by an electromagnetic field with frequency ν and this is the *light absorption*. The light absorption of a conjugated polymer is very important to the efficiency of organic optoelectronic cells.

For an isotropic radiation field, it is necessary to consider contributions from terms of the matrix elements of $\hat{\mu}_y$ and $\hat{\mu}_z$. Since the total radiation, density u is given by the energy per volume per frequency of the incident radiation, the probability of a molecule in the initial state k to perform a transition to the state j when the target is irradiated at the time t_1 is:

and

$$|b_j|^2 = \left(\frac{2\pi t_1}{3\hbar^2}\right)|\langle j|\hat{\mu}|k\rangle|^2 \, u\left(v_{jk}\right) \qquad (5.16)$$

$$v_{jk} = \frac{\omega_{jk}}{2\pi} = \frac{E_j^{(0)} - E_k^{(0)}}{h} \qquad (5.17)$$

where, the electric dipole moment operator is and $\langle j|\hat{\mu}|k\rangle$ is the transition dipole moment. When $\langle j|\hat{\mu}|k\rangle = 0$ the probability of transition vanishes and then this transition is called *electric dipole forbidden*. There are other types of transitions (e.g., magnetic-dipole transitions) but the electric-dipole transitions are the strongest observed in molecular spectroscopy. From the conditions for $\langle j|\hat{\mu}|k\rangle \neq 0$ arise the *selection rules,* which determine the possible transitions.

In the study of electronic transitions in a molecule, the excitation is the promotion of an electron from an orbital of a molecule in the ground state to an unoccupied orbital by absorption of a photon; and this new state is called *excited*. As an example, the π molecular orbital is constituted from p atomic orbitals overlapping. The absorption of a photon of correct energy can change a π-electron to an antibonding orbital π^*. In case of promotion of one of two electrons with opposite spins from the ground state to a molecular orbital of higher energy, in principle, its spin remains the same so that the total spin quantum number is still zero. Due to the multiplicities

($M = 2S + 1$) of both the ground and the excited states, they are named as *singlet state,* and the excitation is a *singlet-singlet* transition. It is also possible a transition in which the promoted electron changes its spin and then, the two electrons with parallel spins yield to a total spin quantum number equal 1 and multiplicity equals 3. This is a *triplet state* once it is correlated with three states of the same energy and has lower energy than the singlet state of the same configuration. Transitions between states of different multiplicities are forbidden. Nevertheless, there is always a weak interaction between the wavefunctions of different multiplicities via a spin-orbit coupling. The crossing from the first singlet excited state to the first triplet state is called intersystem crossing (ISC), which is possible due to spin-orbit coupling.

The excited states of PTs and their decay processes strongly affect the performance of solar cells. After the absorption of a photon by a chain segment, in most of the cases, the excited target relaxes to the lowest vibrational level of the first singlet excited state. Then, it can return to the ground state via several routes. The deactivation may involve a non-radiative decay process, in which the excited-state energy is lost to the environment via vibrational relaxations (VRs) or collisions, or it may involve radiative decay, emission of a photon.

The oscillator strength f is the magnitude of the absolute intensity of the dipole allowed transitions in molecules and f is a dimensionless quantity. It is correlated with the absorption coefficients and the radiation absorption or emission cross-sections. If there are n possible process, $\sum_n f_n$ is equal to the total number of electrons in the target. Assuming that the transition dipoles are oriented at random relative to the direction of the exciting electromagnetic field, the oscillator strength for the transition is (Fox, 2001):

$$f_{jk}(\text{x}) = \frac{2}{3}\frac{m_e}{h^2 e^2}\left(E_k - E_j\right)\mu_{jk}^2 \tag{5.18}$$

where, m_e is the mass of the electron, $\mu_{jk} e \langle k|\hat{r}|j\rangle$ is the transition dipole moment, $\hat{r}$ is the position operator, e is the electronic charge and h is the Planck's constant. The transition moment represents the transient dipole resulting from the displacement of charges during the transition. In solar cells, for example, we seek high oscillator strength at energies where solar irradiance is high.

In PTs, photophysical phenomena are very dependent on distances between segments. The interactions between conjugated parts origin many types of photophysical events which change some polymer optical properties. Some of these polymers can show high optical absorption, and by investigating its backbone structure and conformation, it is possible to correlate the optical absorption with the persistence length of the polymer, for example. The investigation of how to get the ability to tune the magnitude of the optical absorption in PTs could generate many applications, for example, by enabling higher photocurrent generation in photodetectors or solar cells with imperfect charge collection, by increasing the radiative efficiency of solar cells.

5.2.1 FRANCK-CONDON PRINCIPLE

The Franck-Condon principle (Hollas, 2004) also imposes restrictions on excitation processes accessible in molecular spectroscopy. Promotion of an electron to an antibonding molecular orbital upon excitation takes about 10^{-15} s which is very quick compared to the characteristic time for nuclear movement, that is possible to consider the nuclei positions fixed, and the transition between vibrational states is possible just in the situation of the states overlapping. From the solution of the Schrodinger equation in the Born-Oppenheimer approximation, the bond length can be between ranges of probability given by the square of the vibrational wave function for each point. Vertical transitions may occur from any place in this range, according to the nuclei´s instantaneous position at the instant of the excitation. Vertical lines represent vibrational transitions.

Most of the molecules at room temperature are in the ground vibrational state. If during ionization, the bond length is unaffected, the single transition from this ground state of the molecule to the ground state of the ion is more likely, and just one line emerges in the spectrum. The reason for that is due to the probability of each transition between the molecular vibrational level, and the ion vibrational level is proportional to the superposition of the wave functions of the final and initial states.

Once there is no change in the bond length, the overlap integral between the $v_m = 0$ level in the molecule and $v_i = 0$ in the ion assume a high value. However, all the others overlap integrals between the $v_m = 0$ level in the molecule and higher vibrational levels in the ion almost vanish, due to the fact of the positive and negative contributions are canceled. When there is a significant change in the bond length, the highest values of the overlap integrals are the ones regarding excited levels in the ion. Many vibrational levels can be reached in the ion, and a set of lines may be observed in the spectrum. The transition described above was treated according to QM in 1928 by Condon. The intensity of the vibrational transitions is given by:

$$R_{ev} = \int \psi_{ev}^{'*} \mu \psi_{ev}^{'} d\tau_{ev} \qquad (5.19)$$

where, μ is the dipole moment operator and ψ'_{ev} and ψ''_{ev} are the vibrational wave functions of the higher and lower state, respectively. Following the Born-Oppenheimer approximation. This approach allows considering R_e as a constant irrespectively of r, then:

$$R_{ev} = R_e \int \rangle\rangle_V^{'*} \rangle\rangle_V \, dr \qquad (5.20)$$

where, the integral $\int \psi'_v{}^* \psi'_v$ is the overlap integral and its square is the known Franck-Condon factor.

5.2.2   POLYTHIOPHENE (PT) EXCITATIONS

In absorption and fluorescence spectroscopy, there are two crucial molecular orbitals to analyze; the highest occupied molecular orbital (HOMO) and the lowest unoccupied molecular orbital (LUMO), which are related to the ground state of the polymer. The ground state of PTs has electron density localized in the aromatic rings, and the excited state has a quinoidal structure. Regarding band theory of solids, in a polymer, the conduction properties are correlated with its electronic properties. Among them, the band gap and the ionization potential are handy to study these molecules.

The band structures, in particular, the positions of conductions and valence bands and the gap between them, are the main items that specify the inherent properties of the PT. Mainly, the π-electrons play a crucial role in determining the electrical conductivity and band structure in PT. In the PT, the energy of the HOMO-LUMO gap is $E_g = E_{LUMO} - E_{HOMO} = 2.0$ eV (Bundgaard and Krebs, 2007). It is critical to examine the HOMO and the LUMO energies for the oligomers because the relative ordering of the occupied and virtual orbital provides a reasonable qualitative indication of excitation properties (Bouzzine et al., 2015).

5.3 POLYTHIOPHENE (PT) REGIOREGULARITY

The thiophene ring is a 5-membered ring, and the directionality in the polymer is given by polymerization of the monomer through the 2- and 5-position as assigned in Figure 5.2. Whenever a monomer is added to the growing polymer chain, the unit can be bonded in 2-position (Head) or 5-position (Tail) first. Then, the chain is formed by linking ring through its 2 or 5 positions in three relative orientations available: 2,2' or head-to-head (HH) coupling, 5,5' or tail-to-tail (TT) coupling and 2,5' or head-to-tail (HT) coupling, as illustrated in Figure 5.2. Hence, the PT chain that contains only 2,5' or HT coupling is designated RR while a sequence with a mixture of these possible couplings is called regiorandom.

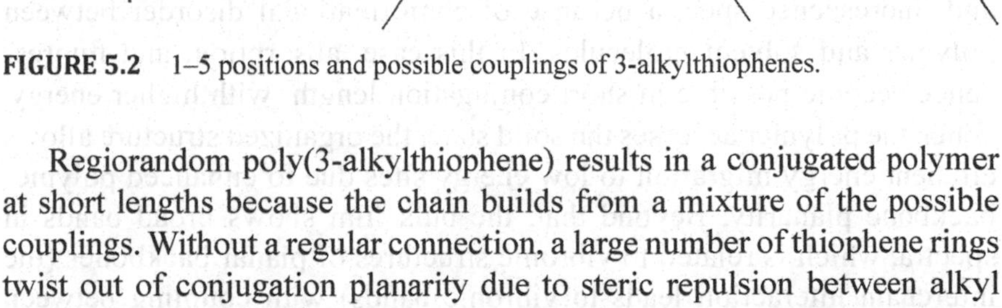

FIGURE 5.2 1–5 positions and possible couplings of 3-alkylthiophenes.

Regiorandom poly(3-alkylthiophene) results in a conjugated polymer at short lengths because the chain builds from a mixture of the possible couplings. Without a regular connection, a large number of thiophene rings twist out of conjugation planarity due to steric repulsion between alkyl chains. Thus, an increase of the torsion angles between thiophene units leads to an increase in the band gaps (Somanathan and Radhakrishnan, 2005). However, RR poly(3-alkylthiophene) can be formed when the coupling between thiophene units occurs in the following H-T configuration, which can adopt a coplanar conformation, resulting in lower energy (Wang et al., 2014). Nevertheless, the planar structure is not enough to eliminate the band gap.

The regioregularity promotes to PT the smallest band gap among the aromatic and hetero-aromatic conjugated polymers. The combination of aromaticity in the ring and a lower energy planar conformation can easily access an extended π-electron system in monomers, leading to highly conjugated polymers (Somanathan and Radhakrishnan, 2005). Studies using femtosecond transient and steady-state spectroscopies determine that primary photoexcitations in RRPT have delocalized polarons and large interchain components. The long-lived photoexcitations are excitons that may separate into polarons among lamellae structure with small relaxation energy. That results in reduced stimulated emission and ISC, giving rise to weak photoluminescence (PL) band. In the contrary, regiorandom PT shows a strong PL band because primary photoexcitations are related to long-lived intrachain polarons with considerable ISC to triplet excitons (Jiang et al., 2002).

Furthermore, a lower energy conformation phenomenon is also observed in absorption and fluorescence spectra when the polymer solution is compared with polymer bulk. As shown in Figure 5.3, the absorption and fluorescence spectra of the polymer bulk film are strongly redshifted (drop to lower band gap) concerning the polymer solution. For films between regiorandom and RR conditions, it should expect the spectrum lying between these repeat units.

It is noticed that polymer solution presents a blue-shift in absorption and fluorescence spectra because of conformational disorder between polymer and solvent molecules. In this case, absorption, and fluorescence become possible in short conjugation length, with higher energy. When the polymer accesses the solid state, the organized structure allows efficient energy migration to low energy sites due to enhanced polymer backbone planarity. Beyond that, the bulk film shows broad bands in spectra, which is related to vibronic structures of planar backbones (the interchain interaction leads to vibronic bands) with coupling between C=C stretching and electronic transition (Clark et al., 2007; Hu et al., 2015).

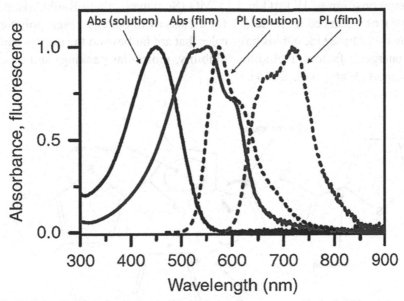

FIGURE 5.3 Typical absorption (solid, Abs) and fluorescence (dashed, PL) spectra of a regioregular poly(3-hexylthiophene) (P3HT) toluene solution and a bulk film.

Source: Reprinted from Hu et al., 2015. Open access.

5.4 POLYTHIOPHENES (PTS) AND SUBSTITUTED POLYTHIOPHENES

Previously, the idea of higher or lower energy transitions was introduced through observation of absorption and fluorescence spectra of a RR polymer. Thereby, the origin of the band gap in PTs can be expressed by the combination of structural contributions, as shown in Figure 5.4, which are intra- and intermolecular interactions, torsion angle between consecutive units, bond length conjugation, resonance effect, and introduction of electron-deficient or electron-sufficient substituent (Wang et al., 2014).

Among these contributions, the nature of the substituent is an essential tool in the elucidation of other factors, since the electronic and structural effects caused by that side groups in thiophene rings decisively tune polymer properties. At electronic and energetic levels, the substituents at side groups can modulate absorption and emission properties and the energy position of HOMO and LUMO (Somanathan and Radhakrishnan, 2005), as observed in Table 5.1 for a class of thiophene-based polymers. However, these side chains have roles that are far beyond the improvement of energetic factors, including solubility, molecular packing, and charge transport (Wang et al., 2014).

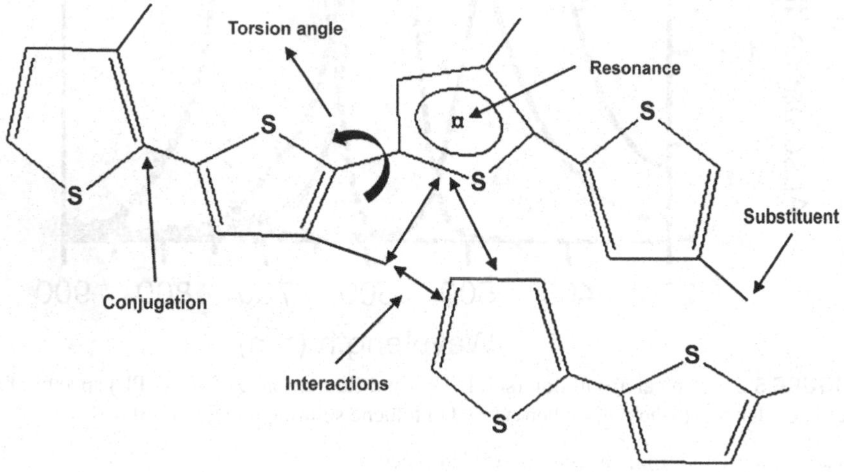

FIGURE 5.4   Structural contributions for polythiophene polymers.
Source: Adapted from Wang et al., 2014.

The arrangement of the monomer units in substituted PTs especially with bulky substituents can modify its conformational features, which in turn, govern the degree of π-π conjugation between adjacent rings. While PTs functionalized with conjugated pendant groups possess broad absorption band in the ultraviolet and visible regions (UV/Vis) (Wang et al., 2014), increasing the alkyl chain promotes an increase of conjugation length, resulting in a decreasing of band gap at electronic properties in the substituted PT (Somanathan and Radhakrishnan, 2005). This behavior can be noted in Table 5.1 comparing the maximum in the absorption spectrum (λ_{abs}) and the HOMO and LUMO energy levels. Besides, differences between λ_{abs} from polymer solution and film could be gathered for some polymers, which are observed a redshift in the absorption spectrum from solution to film. Then, the solid-state generally shows a lower band gap related to increased interaction between the chains.

TABLE 5.1 Energetic Parameters from Thiophene-Based Polymers

Polymer		λ_{abs} (nm)	HOMO (eV)	LUMO (eV)	Band-Gap (eV)
Polythiophene[a]			−4.26	−3.21	1.05
Polythiophene[b]	Film	552	−4.76	−2.74	2.02
Poly(methylthiophene)[a]	Film	506	−3.94	−2.87	1.07
Poly(3-ethylthiophene)[a]	Film	460			
Poly(3-butylthiopene)[a]	Solution	434			
Poly(3-isobutylthiophene[a]	—	426			
Poly(3-hexylthiophene) (P3HT)[a]	Solution	434	−3.79	−2.80	0.99
	Film	504			
Poly(3-hexylthiophene) (P3HT)[c]	Solution	439			1.91
	Film	510			
Poly(3-octylthiophene) (P3OT)[d]	Solution	448			2.3
	Film	514			1.9
Poly(3-dodecylthiophene)[a]	Film	521			
polythiophenes with bi(thienylenevinylene) side groups[b] P1	Film	413	−4.96	−2.97	1.99
polythiophenes with bi(thienylenevinylene) side groups[b]P2		445	−4.94	−2.95	1.99
polythiophenes with bi(thienylenevinylene) side groups[b]P3		452	−4.93	−2.96	1.97
Polythiophenes with cyano side groups[d] PBCN4HT	Solution	369			2.7
	Film	372			2.5
Polythiophenes with cyano side groups[d] P3CN4HT	Solution	392	−6.1	−3.6	2.5
	Film	408			2.3

[a] Data from Samsonidze et al. (2014). DFT calculations for isolated polymers.
[b] Data from Hou et al. (2006). Energy levels calculations from cyclic voltammetry and optical band gap.
[c] Data from Chang et al. (2008). Optical band gap.
[d] Data from Chochos et al. (2007). Energy levels calculations from cyclic voltammetry and optical band gap.

In 2006, Hou et al. first designed and synthesized a series of PT deriva-tives having conjugated bi(thienylenevinylene) side chains (Figure 5.5). It has been shown that the extension of a conjugated side chain to the PT main chain leads to broad and robust absorption covering both the UV/Vis from 350 to 650 nm. The ultraviolet band results from the absorption of the bi(thienylenevinylene) side chains (350–450 nm), and the visible peak is attributed to the absorption of the π–π^* transition of the conjugated polymer main chains (450–650 nm) (Hou, 2006).

Yu et al. developed RR PTs possessing alkyl-thiophene side chains which the insertion of two thiophene conjugated side chains was able to lower bandgap to 1.77 eV (Yu et al., 2009), compared with experimental data of non-substituted PTs. If the side chain is also an electron-donating group, it can contribute some degree of electron density to the conjugated polymer backbones. The incorporation of side groups as triphenylamine (TPA) has been included in PT to promote broader peaks in the visible absorption and lower HOMO energy level when compared to alkyl chain side in P3HT (Li et al., 2008). Conjugated side groups can lower HOMO levels generally attributed to the conjugation amplification along the polymer backbone (Wang et al., 2014).

Another approach to manipulating the energy levels of PT is the introduc-tion of an electron acceptor group as the pendant on the polymer backbone. The insertion of an electron acceptor was able to lower both HOMO and LUMO levels and increase band gap as observed to the cyano group (NC) in the two polymers showed at Figure 5.5 (Chochos et al., 2007).

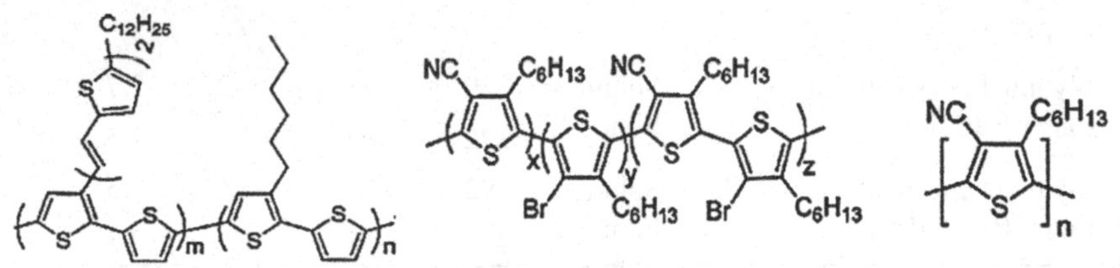

FIGURE 5.5   Bi(thienylenevinylene) substituted polythiophene and two cyano substituted polythiophene, respectively.

5.4.1   POLY(3-HEXYLTHIOPHENE) (P3HT)

Among a big family of PTs, poly(3-hexylthiophene) (P3HT) is the most commonly used material due to many advantages such as easy synthesis, good processability and high charge carrier mobility (Wang et al., 2014). The ground state of P3HT has electron density localized in the heterocycle, related to a noncoplanar structure. However, the excited state has a quinoidal framework (Barnes and Baghar, 2012) and as suggested by Spano (2010; Spano et al., 2009), P3HT has an organization frequently discussed as H- and J-type aggregates within the framework of Frenkel polaronic excitons. These aggregates distinguish in H-type where the thiophene rings are noncoplanar, while in J-type thiophene rings adopt a high-degree of planarization. As a result, excitonic coupling in interchain is observed in H-aggregates, and intrachain is denoted in J-aggregates (Barnes and Baghar, 2012). It is noted on the absorption spectrum that P3HT film has a broader band when compared with P3HT solution. This wide band can be understood as vibronic transitions labeled as 0-n (n = 0, 1, 2,…) where n is the number of vibrational quanta in the final excited state.

The photophysical properties of P3HT and related compounds reveal a 2D electronic delocalization due to a high level of interchain ordering in lamellae structures. That organization promotes a decrease in fluorescence quantum yield; however, an improvement in fluorescence rate is observed and followed by an enhanced nonradiative decay rate (Thiessen et al., 2013). If charge-transfer (CT) process occurs between different polymer phases, these CT-type states are located at the interface ordered/disordered phase driven by energetic disorder (Glowe et al., 2010; Paquin et al., 2011).

5.5 MONOMERS CONTAINING THIOPHENE UNIT AND THIOPHENE-BASED COPOLYMERS

In recent years, monomers which combine thiophene ring and other units as benzothiadiazole (BT) and benzodithiophene (BDT) groups have been widely studied due to structural symmetry of the rigid fused aromatic system. These groups in the polymer chain, as shown in Figure 5.6, enables an improvement in electron delocalization, resulting in conjugated polymers with a low energy two-dimensional structure.

Additional inclusion of thiophene units as side chain was proposed by Huo et al. (2011) in order to improve BDT-based polymers. The synthesized polymer PBDTTBT (Figure 5.6) presented lower band gap and HOMO/LUMO levels compared with PT (Table 5.2). The absorption spectrum revealed a wide absorption band between 300 and 700 nm, with λ_{max} at ~581 nm in chloroform and 596 nm in the film (Hou et al., 2010).

5-Alkylthiophene-2-yl-substituted BDT were synthesized to result in PBDTT-based polymers (Figure 5.6). For these polymers, the enhanced intermolecular π–π interaction of the two-dimensional packing results in a redshifted and broadened absorption band when compared to those of their analogous polymers without alkylthienyl substituents (Wang et al., 2014; Huo et al., 2011). Another BDT-based polymer with thiophene as a pendant group was synthesized by Zhang et al. and resulted in a low bandgap polymer. PBDTT-BT (Figure 5.6) showed an optical band gap of 1.67 eV (Zhang et al., 2013), lower than PBDTTBT. Using thiophene units only as π-bridge in PBDTF-BT, a redshifted absorption band is also obtained, and a consequent low band gap of 1.70 eV is observed for the polymer (Zhang et al., 2013). Fluorination of BT ring in PBDTT-based polymers further reduces the HOMO energy due to interactions of F atom in the intra- and intermolecular levels (Zhou et al., 2011).

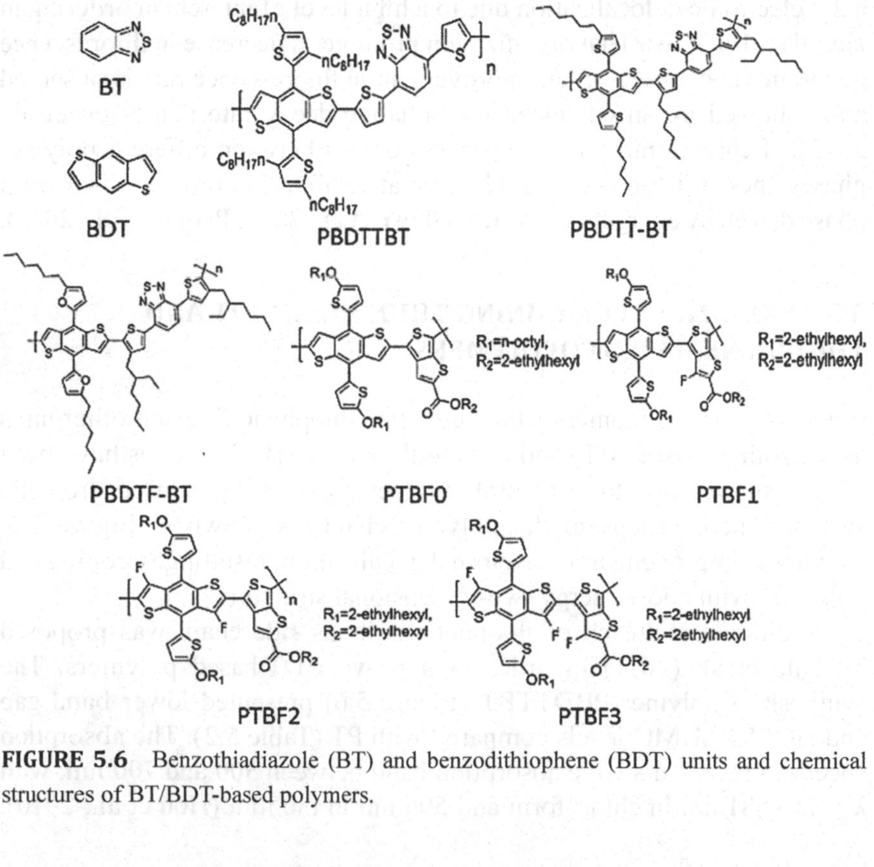

FIGURE 5.6   Benzothiadiazole (BT) and benzodithiophene (BDT) units and chemical structures of BT/BDT-based polymers.

Fluorination of the thiophene ring in the polymer backbone is also capable of lowering HOMO and LUMO levels. Son et al. have synthesized a class of fluorinated polymers with similar structures (PTBF0, PTBF1, PTBF2, and PTBF3) as observed in Figure 5.6. The absorption profiles did not change after fluorination; however, the PTBF-based polymers showed a strong influence on lowering HOMO levels (Table 5.2) (Son et al., 2011).

Wei and co-workers reported a series of RR 3-hexylthiophene copolymers incorporated with different composition ratios of octyl-phenanthrenyl-imidazole (Figure 5.7). They related that with 90 mol% octylphenanthrenyl-imidazole moieties onto P3HT chains, the band gap was effectively reduced to 1.80 eV when compared with P3HT. This fact was attributed to the increased effective conjugation length of the PT main chain (Chang et al., 2008).

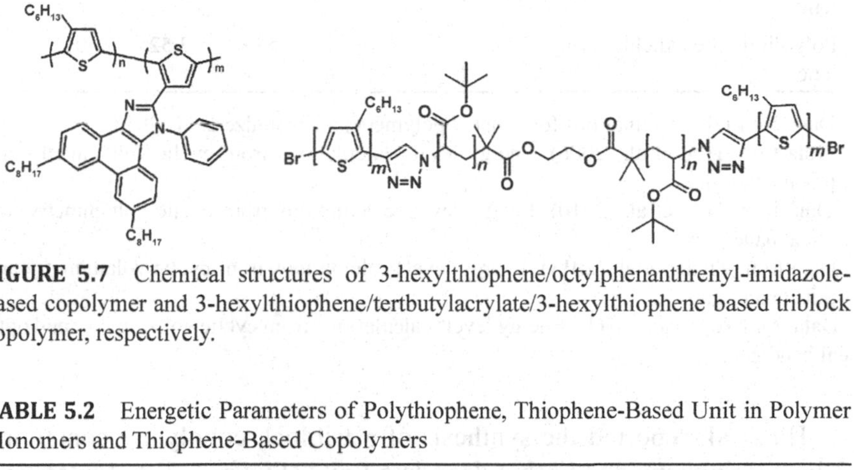

FIGURE 5.7 Chemical structures of 3-hexylthiophene/octylphenanthrenyl-imidazole-based copolymer and 3-hexylthiophene/tertbutylacrylate/3-hexylthiophene based triblock copolymer, respectively.

TABLE 5.2 Energetic Parameters of Polythiophene, Thiophene-Based Unit in Polymer Monomers and Thiophene-Based Copolymers

Polymer		λ_{abs} (nm)	HOMO (eV)	LUMO (eV)	Band-Gap (eV)
Polythiophene[a]			−4.26	−3.21	1.05
Polythiophene[b]	Film	552	−4.76	−2.74	2.02
PBDTTBT[c]	Solution	581	−5.31	−3.44	1.75
	Film	596			
PBDTT-BT[d]	Solution	548	−5.26	−3.34	1.67
	Film	604			
PBDTF-BT[d]	Solution	566	−5.24	−3.54	1.70
	Film	608			

TABLE 5.2   *(Continued)*

Polymer		λ_{abs} (nm)	HOMO (eV)	LUMO (eV)	Band-Gap (eV)
PTBF0[e]	Film	683	−4.94	−3.22	1.59
PTBF1[e]	Film	671	−5.15	−3.31	1.68
PTBF2[e]	Film	670	−5.41	−3.60	1.75
PTBF3[e]	Film	670	−5.48	−3.59	1.73
Poly(dithiophene ethyne)[a]			−4.44	−3.36	1.08
Poly(dithiophene vinylene)[a]			−4.19	−3.36	0.83
Poly(3-hexyldithiophene vinylene)[a]			−3.81	−3.12	0.69
Poly(dithiophene dicyanovinylene)[a]			−5.26	−4.63	0.63
Poly(dithiophenedifluoroviny lene)[a]			−4.43	−3.46	0.97
Poly(dithiophenedichloroviny lene)[a]			−4.53	−3.52	1.01

[a] Data from DFT calculations for isolated polymers (Samsonidze et al., 2014).
[b] Data from Hou et al. (2006). Energy levels calculations from cyclic voltammetry and optical band gap.
[c] Data from Hou et al. (2010). Energy levels calculations from cyclic voltammetry and optical band gap.
[d] Data from Zhang et al. (2013). Energy levels calculations from cyclic voltammetry and optical band gap.
[e] Data from Son et al. (2011). Energy levels calculations from cyclic voltammetry and optical band gap.

Hu et al. reported the synthesis of a triblock copolymer containing 3-hexylthiophene and tertbutylacrylate units (Figure 5.7) to understand the nature and properties of excited states in the PT derivative via aggregates suspended in solvents. First, they applied fluorescence correlation spectroscopy (FCS) to examine the variation in several emitters upon going from toluene to toluene/poor solvent mixture. The results revealed that when going from good (toluene) to poor solvent (toluene/methanol mixture), the number of emitters approximately drops 5–10 times. This result reveals that triblock forms aggregates in a poor solvent, but single chains merely collapse and do not aggregate. In absorption spectra, as observed previously with an increase of poor solvent, a redshifted spectrum with a rise in gradual intensity of vibronic structures due to interchain interaction was observed. In fluorescence measurements, the authors obtained spectra under excitation at 560 nm, at which only aggregated P3HT can be excited. These spectra exhibit bulk film-like spectral profile but with slight variation in vibronic structure. However, in the fluorescence lifetime, they observed similarities in the lifetime of the emissive excitonic state of the triblock in both polar and nonpolar solvents. At that point, there is a branching between the generation of exciton-type state and CT-type state. That is, the CT-type state exists as a competing channel of the formation of an exciton-type state (Hu et al., 2015).

5.6 CONCLUDING REMARKS

This chapter highlights the role of conjugated PT and its derivatives and how they are capable of tuning photophysical characteristics including absorption, emission, and energy levels. Depending on the application, these parameters are critical to obtaining a polymer with that desired properties. For example, the performance of solar cells is directly affected by excited states in PT and their decay processes. If the polymer segments absorb a photon, the excited-state must involve non-radiative and/or radiative decay processes, which are strongly affected by polymer backbone and polymer bulk structure. Transient and steady-state studies show delocalized polarons and large interchain components with little relaxation energy and weak PL band in RRPT. On the contrary, a strong PL band is observed in regiorandom PT due to long-lived intrachain polarons with considerable ISC to triplet excitons.

From the substituent type in PT derivatives, some structural characteristics have been noticed from the literature. First, the insertion of alkyl and conjugated side groups were associated with a decrease of the HOMO level. However, it is not directly interrelated with a lower band gap. Second, the size and conformation of the side group also influence polymer energetics because of its molecular volume and packing, and both properties impact directly at the main chain polymer framework. A good side group allows a more planar backbone along thiophene rings and can lower band gap due to effective π–π stacking. And finally, polymers with rigid fused units like BT and BDT are capable of lowering HOMO level as well as the incorporation of an electron or donor groups, both owing to electron delocalization along thiophene ring and substituent and/or thiophene coplanar structures.

KEYWORDS

- fluorescence correlation spectroscopy
- highest occupied molecular orbital
- polythiophene
- polythiophene derivatives
- substituted polythiophenes
- thiophene

REFERENCES

Barnes, M. D., & Baghar, M., (2012). Optical probes of chain packing structure and exciton dynamics in polythiophene films, composites, and nanostructures. *J. Polym. Sci. Part B Polym. Phys., 50*, 1121–1129.

Bolinger, J. C., Traub, M. C., Brazard, J., Adachi, T., Barbara, P. F., & Vanden, B. D. A., (2012). Conformation and energy transfer in single conjugated polymers. *Acc. Chem. Res., 45*(11), 1992–2001.

Bouzzine, S. M.,Salgado-Morán, G., Hamidi, M., Bouachrine, M., Pacheco, A. G., & Glossman-Mitnik, D., (2015). DFT study of polythiophene energy band gap and substitution effects. *J. Chem., 2015*, 296386.

Bundgaard, E., & Krebs, F. C., (2007). Low band gap polymers for organic photovoltaics. *Sol. Energy Mater. Sol. Cells, 91*(11), 954–985.

Chang, Y. T., Hsu, S. L., Chen, G. Y., Su, M. H., Singh, T. A., Diau, E. W. G., & Wei, K. H., (2008). Intramolecular donor-acceptor regioregular poly (3-hexylthiophene)s presenting octylphenanthrenyl-imidazole moieties exhibit enhanced charge transfer for hetero junction solar cell applications. *Adv. Func. Mater., 18*, 2356–2365.

Chochos, C. L., Economopoulos, S. P., Deimede, V., Gregoriou, V. G., Lloyd, M. T., Malliaras, G. G., & Kallitsis, J. K., (2007). Synthesis of a soluble n-type cyano substituted polythiophene derivative: A potential electron acceptor in polymeric solar cells. *J. Phys. Chem. C, 111*, 10732–10740,

Clark, J., Silva, C., Friend, R. H., & Spano, F. C., (2007). Role of intermolecular coupling in the photophysics of disordered organic semiconductors: Aggregate emission in regioregular polythiophene. *Phys. Rev. Lett., 98*(20), 206406.

Cohen-Tannoudji, C., Diu, B., & Laloe, F., (1992). *Quantum Mechanics* (p. 2). Hermann.

De Melo, J. S., Burrows, H. D., Svensson, M., Andersson, M. R., & Monkman, A. P., (2003). Photophysics of thiophene based polymers in solution: The role of nonradiative decay processes. *J. Chem. Phys., 118*(3), 1550–1556.

De Melo, J. S., Silva, L. M., Arnaut, L. G., & Becker, R. S., (1999). Singlet and triplet energies of alpha-oligothiophenes: A spectroscopic, theoretical, and photoacoustic study: Extrapolation to polythde. *J. Chem. Phys., 111*(12), 5427–5433.

Fox, M., (2001). *Optical Properties of Solids* (1st edn.). Oxford master series in condensed matter physics; Oxford University Press.

Glowe, J. F., Perrin, M., Beljonne, D., Hayes, S. C., Gardebien, F., & Silva, C., (2010). Charge-transfer excitons in strongly coupled organic semiconductors. *Phys. Rev. B, 81*, 041201.

Hollas, J. M., (2004). *Modern Spectroscopy* (4th edn.). John Wiley and Sons.

Hou, J., Tan, Z. A., Yan, Y., He, Y., Yang, C., & Li, Y., (2006). Synthesis and photovoltaic properties of two-dimensional conjugated polythiophenes with bi(thienylenevinylene) side chains. *J. Am Chem. Soc., 128*, 4911–4916.

Hu, Z., Willard, A. P., Ono, R. J., Bielawski1, C. W., Rossky, P. J., & Bout, D. A. V., (2015). An insight into non-emissive excited states in conjugated polymers. *Nat. Commun., 6*, 8246.

Huo, L., Guo, X., Li, Y., & Hou, J., (2011). Synthesis of a polythieno [3, 4-b] thiophene derivative with a low-lying HOMO level and its application in polymer solar cells. *Chem. Commun., 47*, 8850–8852.

Huo, L., Hou, J., Zhang, S., Chen, H. Y., & Yang, Y., (2010). A polybenzo [1,2-b:4,5-b′] dithiophene derivative with deep HOMO level and its application in high-performance polymer solar cells. *Angew. Chem. Int. Ed., 49*, 1500–1503.

Jen, K. Y., Miller, G. G., & Elsenbaumer, R. L., (1986). Highly conducting, soluble, and environmentally-stable poly(3-alkylthiophenes). *J. Chem. Soc., Chem. Commun.*, 1346–1347.

Jiang, X. M., Österbacka, R., Korovyanko, O., An, C. P., Horovitz, B., Janssen, R. A. J., & Vardeny, Z. V., (2002). Spectroscopic studies of photoexcitations in regioregular and regiorandom polythiophene films. *Adv. Func. Mater., 12*(9), 587–597.

Kuhn, H., (1949). A quantum-mechanical theory of light absorption of organic dyes and similar compounds. *J. Chem. Phys., 17*(12), 1198.

Levine, I. N., (1975). *Molecular Spectroscopy* (2nd edn.). John Wiley and Sons.

Li, Y., Xue, L., Xia, H., Xu, B., Wen, S., & Tian, W., (2008). Synthesis and properties of polythiophene derivatives containing triphenylamine moiety and their photovoltaic applications. *J. Polym. Sci. A Polym. Chem., 46*, 3970–3984.

McNeill, C. R., Abrusci, A., Hwang, I., Ruderer, M. A., Müller-Buschbaum, P., & Greenham, N. C., (2009). Photophysics and photocurrent generation in polythiophene/ polyfluorene copolymer blends. *Adv. Funct. Mater., 19*(19), 3103–3111.

Paquin, F., Latini, G., Sakowicz, M., Karsenti, P. L., Wang, L., Beljonne, D., Stingelin, N., & Silva, C., (2011). Charge separation in semi crystalline polymeric semiconductors by photo excitation: Is the mechanism intrinsic or extrinsic? *Phys. Rev. Lett., 106*, 197401.

Patil, A. O., Heeger, A. J., & Wudl, F., (1988). Optical properties of conducting polymers. *Chem. Rev., 88*(1), 183–200.

Samsonidze, G., Ribeiro, F. J., Cohen, M. L., & Louie, S. G., (2014). Quasiparticle and optical properties of polythiophene-derived polymers. *Phys. Rev. B, 90*, 035123.

Sato, M., Tanaka, S., & Kaeriyama, K., (1986). Soluble conducting polythiophenes. *J. Chem. Soc., Chem. Commun., 873–874.*

Somanathan, N., Radhakrishnan, S., (2005). Optical properties of functionalized polythiophenes. *Int. J. Mod. Phys. B., 19*(32) 4645–4676.

Son, H. J., Wang, W., Xu, T., Liang, Y., Wu, Y., Li, G., & Yu, L., (2011). Synthesis of fluorinated polythienothiophene-co-benzodithiophenes and effect of fluorination on the photovoltaic properties. *J. Am. Chem. Soc., 133,* 1885–1894.

Spano, F. C., (2010). The spectral signatures of Frenkel-Polarons in H-and J-aggregates. *Acc. Chem. Res., 43,* 429–439.

Spano, F. C., Clark, J., Silva, C., & Friend, R. H., (2009). Determining exciton coherence from the photoluminescence spectral line shape in poly(3-hexylthiophene) thin films. *J. Chem. Phys., 130*(7), 074904–074920.

Springborg, M., (1992). The electronic properties of polythiophene. *J. Php. Condens. Matter, 4,* 101–120.

Thiessen, A., Vogelsang, J., Adachib, T., Steiner, F., Bout, D. V., & Lupton, J. M., (2013). Unraveling the chromophoric disorder of poly(3-hexylthiophene). *PNAS, 110*(38) E3550–E3556.

Vezie, M. S., Few, S., Meager, I., Pieridou, G., Dörling, B., Ashraf, R. S., Goñi, A. R., et al., (2016). Exploring the origin of high optical absorption in conjugated polymers. *Nat. Mater., 15*(7), 746–753.

Wang, H. J., Chen, C. P., & Jeng, R. J., (2014). Polythiophenes comprising conjugated pendants for polymer solar cells: A review. *Materials, 7*(4), 2411–2439.

Yamamoto, T., (2010). Molecular assembly and properties of polythiophenes. *NPG Asia Mater,2,* 54–60.

Yu, C. Y., Ko, B. T., Ting, C., & Chen, C. P., (2009). Two-dimensional regioregular polythiophenes with conjugated side chains for use in organic solar cells. *Sol. Energy Mater. Sol. Cells, 93,* 613–620,

Zhang, Y., Gao, L., He, C., Sun, Q., & Li, Y., (2013). Synthesis and photovoltaic properties of two-dimension-conjugated D–A copolymers based on benzodithiophene or benzodifuran units. *Polym. Chem., 4,* 1474–1481.

Zhou, H., Yang, L., Stuart, A. C., Price, S. C., Liu, S., & You, W., (2011). Development of fluorinated benzothiadiazole as a structural unit for a polymer solar cell of 7% efficiency. *Angew. Chem. Int. Ed., 50,* 2995–2998.

CHAPTER 6

Photophysical Properties of Polyfluorenes

GUSTAVO T. VALENTE[1] and NIRTON C. S. VIEIRA[2]

[1]*São Carlos Institute of Physics, University of São Paulo, PO Box – 369, 13560-970, São Carlos, SP, Brazil, E-mail: gtvfisica@gmail.com*

[2]*Institute of Science and Technology, Federal University of São Paulo, 12231-280, São José dos Campos, SP, Brazil*

6.1 INTRODUCTION

Polyfluorenes (PFO) are a class of conjugated polymers which show a typical blue luminescence in solution and solid-state. Fluorene molecules form a polymer backbone, as shown in Figure 6.1(a). These polymers exhibit interesting physical properties, as photoluminescence (PL) quantum efficiency can rise above 50%, and charge mobility can be controlled by molecular morphology (Ariu et al., 2002; Perevedentsev et al., 2016; Prins et al., 2006). Therefore, they are attractive for applications in optoelectronic industry, mainly for developing organic electronic devices, such as light-emitting diodes (Peet et al., 2008) and solar cells (Oh et al., 2010).

In luminescence-based sensors, PFO have been combined with analyte-sensitive materials for the detection of chemical compounds. In recent reports, polyfluorene functionalized with terpyridine and truxene were used for fluorescence sensing of the Cu^{2+} and Fe^{2+} ions (Juang et al., 2016; Yang et al., 2016). Detection limit with the order of magnitude of 10^{-6} and 10^{-7} M^{-1} were observed for the detection of Cu^{2+} and Fe^{2+} ions, respectively. These values are compared to well-established electrochemical techniques (Bansod et al., 2017), showing the potential of these polymers for monitoring ions in environmental samples (Juang et al., 2016; Yang et al., 2016). In the same way, *HCl* gas sensor via Förster transfer (dipole-dipole coupling) using thin films of polyfluorene doped with brocomocresol green has also been developed (Guillén et al., 2016). The Förster transfer is the physical mechanism of the exciton (photogenerated electron-hole pair) migration in conjugated polymer films that are directly related to the molecular morphology and conformation disorder of conjugated polymer (Forster, 1959).

PFO can exhibit different molecular morphology such as amorphous (also so-called glassy), β (noncrystalline) and crystalline (α and α') phases while keeping the same chemical composition (Chen, Su, and Chen, 2005; Grell et al., 1999). Polyfluorene films with β-phase chain dispersed within an amorphous matrix are particularly interesting for photophysical studies due to the energy transfer between these phases (Ariu et al., 2003). Grell et al. (1999) proposed molecular structural model named first β-phase to explain the spectroscopic properties in polyfluorene films. In this molecular model, the rotation angle between adjacent monomer units is 180°, forming a highly ordered chain of the β-phase, as exemplified by Figure 6.1(b) for poly(9,9-dioctylfluorene) (PFO). Its conjugation length has around 30 monomer units. Different physicochemical treatments, to generate the β-phase chain in amorphous films have been reported such as exposure toluene vapor, heating-cooling cycles from room temperature to cryogenic temperatures, and the use of poor solvents (Grell et al., 1999).

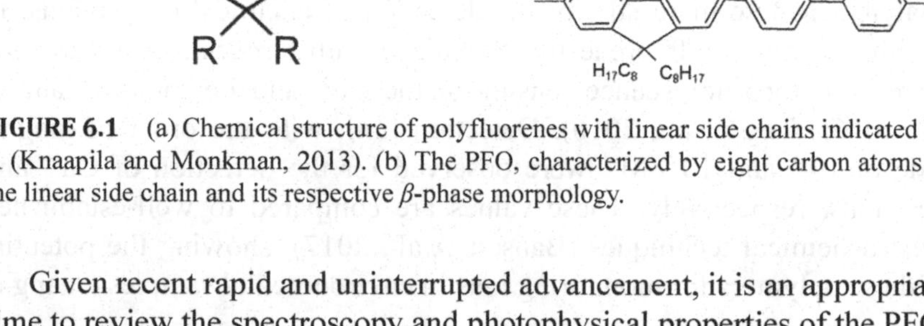

FIGURE 6.1   (a) Chemical structure of polyfluorenes with linear side chains indicated by R (Knaapila and Monkman, 2013). (b) The PFO, characterized by eight carbon atoms in the linear side chain and its respective β-phase morphology.

Given recent rapid and uninterrupted advancement, it is an appropriate time to review the spectroscopy and photophysical properties of the PFO.

The relation between phase morphological and photophysical properties for PFO films with particular attention to the exciton migration are discussed in this chapter. The organization of this chapter is as follows. Section 6.2 describes how the conformational disorder in a conjugated polymer is vital on its spectroscopic and photophysical properties. Optical features of PFO are presented in Section 6.3. The exciton migration in PFO is covered in Section 6.4 with particular attention to molecular morphology effect and concluding remarks are given in Section 6.5.

6.2 DISORDERED CONJUGATED POLYMER: A GOOD PICTURE OF POLYFLUORENE

Conformational disorder in the conjugated polymer plays an essential role in its spectroscopic and photophysical properties. The critical concept is based on the energetic disorder produced by the conformational disorder. A simple description includes the energy gap (Eg) dependence over polymeric chain length with the conformational distribution.

Figure 6.2 shows the E_g calculated for different chain lengths of oligofluorenes (the chemical structure is shown in the inset of Figure 6.2), where, n represents the monomer units (Choi et al., 2014). E_g increases with the inverse of the chain lengths ($1/n$). Using the exciton theory (Beenken; Pullerits, 2004), E_g is described as follows:

$$E_g(n) = E_1 - E_2 \cos\left(\frac{\pi}{n+1}\right)$$

(6.1)

where, the values of E_1 = 4.86 eV and E_2 = 1.41 eV were obtained from curve fitting in Figure 6.2.

According to the random distribution for different chain subunit lengths and the $E_g(n)$, the density of state (DOS) can be obtained. In this approach, each polymeric chain subunit corresponds to a localized state in which the exciton occupies. For example, assuming the Gaussian distribution of the chain length as:

$$g(n) = \frac{1}{\sqrt{2\pi\sigma^2}} e^{-\frac{(n-\mu)^2}{2\sigma^2}}$$

(6.2)

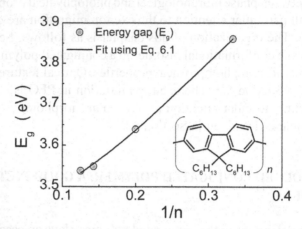

FIGURE 6.2 Energy gap vs. inverse of the number of monomers ($1/n$) calculated by Choi and co-workers (2014) for oligofluorene with chemical structure inserted in the figure. Adapted from Choi and co-workers (2014) and fitted using the exciton theory (Eq. (6.1)) from Beenken and Pullerits (2004).

where, σ^2 is the variance and μ the mean length, the energies distribution is obtained by rewriting it as following:

$$g\left(E_g\right)=\frac{1}{\sqrt{2\pi\sigma^2}}e^{-\frac{1}{2\sigma^2}\left(\frac{\pi}{\arccos\left(\frac{E_1-E_g}{E_2}\right)}-1-\mu\right)^2} \tag{6.3}$$

where, $n\left(E_g\right)=\dfrac{\pi}{\arccos\left(\dfrac{E_1-E_g}{E_2}\right)}-1$ from Eq. (6.1). This equation

describes the DOS of localized states produced by the Gaussian distribution of the chain length.

In contrast to the chain length distribution, the DOS shows an asymmetric behavior with approximate Gaussian DOS (hatched area in Figure 6.3(b)) agreeing very well for low energies. Gaussian DOS is commonly used and powerful to describe exciton dynamics and charge transport in disordered organic materials (Laquai et al., 2009; Nenashev, Oelerich, and Baranovskii, 2015).

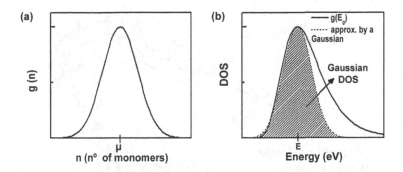

FIGURE 6.3 Gaussian distribution of the chain length from Eq. (6.2), (b) DOS obtained using the energy distribution given by Eq. (6.2) and approximate Gaussian DOS (hatched area).

Hu and co-workers (2000) combined computer simulation with polymerization spectroscopy measurements to study the conformational properties of poly(2-methoxy-5-29-ethylhexyloxy-1,4-phenylenevinylene) (MEH-PPV) in a single-molecule approach. They found that 'tetrahedral chemical defects' promote subunits in MEH-PPV chain, that adopts a cylindrical conformational. From this point of view, conformational disorder along polymeric chain disrupts the π-electron conjugation (conjugation break) promoting conformation subunits that act as chromophores. The authors suggest the same conformational feature as a candidate model to describe PFO chains.

Another mechanism to produce conformational disorder is the torsion angle in the polymeric chain due to thermal effect. Alternatively, Scholes (2006) points to the difficulty of relating chain subunits generation to the torsional chain. The author highlights that even the weak π-π overlap must be considered for the understanding of excitons in conjugated polymers. Quantum chemical calculation about segmentation mechanism in conjugated polymer reveals that geometric defect (like torsion, for example) does not produce exciton localization in specific chain subunits (Beenken and Pullerits, 2004). The consideration of electronic coupling between the chain subunits ensures the absorption feature in the conjugated polymers via photogeneration of delocalized excitons (Yang, Dykstra, and Scholes, 2005). A simple picture to describe the conformational disorder is shown in Figure 6.4 for PFO.

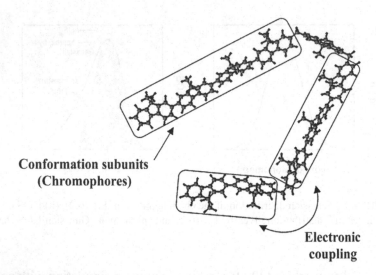

**Conformation subunits
(Chromophores)**

**Electronic
coupling**

FIGURE 6.4 Part of a polyfluorenes chain with conformation subunits based in the chemical-physical defect. The disrupt π-electron conjugation promoting conformation subunits (chromophores) over the polyfluorenes chain. However, the subunits next to each other are electronic coupling ensuring the energy transfer of the exciton between them. Based on the illustration reported by (Scholes and Rumbles, 2006) for poly(phenylene vinylene).

In this context, the standard definition of chromophores as "well-defined subunits of the polymeric chain that absorb and emit light" needs to be revised. The trilogy of papers reported by Barford and co-workers show advancement in the understanding of the chromophores based in the concept of local exciton ground states (LEGSs), also named as LGS (local ground state) (Barford and Marcus, 2014; Barford and Tozer, 2014; Marcus, Tozer, and Barford, 2014). LEGSs are characterized by the spatially localized and nonoverlapping lowest energy excited states (Makhov and Barford, 2010). These features can define chromophores. However, a local exciton ground state may contain one more chain segment delimited by the conjugation break when the exciton localization length exceeds the length scale between the conjugation breaks (Marcus, Tozern, and Barford, 2014).

Regardless of the chromophores definition, there is the energy distribution coming from conformational disorder. This feature is enough to explain the absorption and emission properties as well as the exciton migration in conjugated polymers such as PFO.

6.3   OPTICAL FEATURES OF POLYFLUORENES (PFO)

Energetic disorder produced by conformational disorder influences directly on the absorption properties of PFO. Figure 6.5(a) shows the absorption spectrum at room temperature of amorphous PFO film produced by spin-coating technique (Ariu et al., 2003). PFO absorption is characterized by blue broadband spectrum. It originates from electronic transition from ground state S_0 to first excited electronic state S_1 along with vibrational shift for entire chromophores. In contrast, PFO PL exhibits emission at 423, 447, 476, and 502 nm as shown in Figure 6.5(a). The higher intensity emission (423 nm) results from the purely electronic transition (0–0 transition) followed by 0–1, 0–2, 0–3 vibronic transition at 447, 476, and 502 nm, respectively. Amorphous PFO films show PL quantum efficiency (PLQE) around 53% (Perevedentsev et al., 2016; Ariu et al., 2002).

At the temperature of 5 K, PL spectrum is well resolved with a redshifted emission (428, 456, 486, and 515 nm) while the absorption spectrum is not affected as shown in Figure 6.5(b) (Ariu et al., 2003). Regardless of the temperature, the whole PL spectrum is redshifted about the main band absorption. This feature results in spectral diffusion, promoted by exciton migration for the lower energy states in the DOS tail, which are responsible for the PL. Detail of the exciton migration and PL relationship is discussed in the next section.

The intensities of the vibronic transitions are evaluated from the Franck-Condon factor (F_{0v}). This factor provides the weight of each vibronic transition (Barford, 2005) as follows:

$$F_{0v} = \frac{S^v}{v!} e^{-S}$$

$$(6.4)$$

where, S is the Huang-Rhys parameter and v the vibrational level ($v = 0$, 1, 2, 3, 4...). From the intensities of 0–0 (I_{0-0}) and 0–1 (I_{0-1}) transitions in the experimental PL spectrum. The Huang-Rhys parameter is obtained, which is expressed by $S = I_{0-1}/I_{0-0}$ (Barford, 2005). Figure 6.6 shows the calculated F_{0v} against v expected for amorphous PFO film for $S = 0.5$ estimated from the PL spectrum at 5 K, of Figure 6.5(b). Moreover, above $v = 3$ (0–3 transition), F_{0v} assumes low values, and consequently, only three vibronic transitions are observed, experimentally.

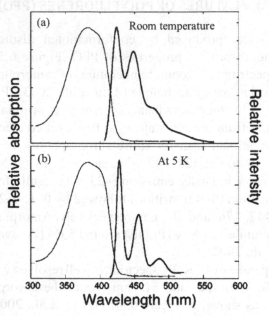

FIGURE 6.5   Absorption and photoluminescence spectrum of amorphous PFO film at (a) room temperature and (b) at 5 K. The sample was excited using HeCd laser at 354 nm in the photoluminescence measurement.

Source: Reprinted with permission from Ariu et al. (2003). Copyright (2003) by the American Physical Society.

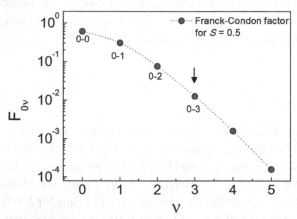

FIGURE 6.6   Franck-Condon factor as a function of v for $S = 0.5$ for amorphous PFO film. S was estimated from the photoluminescence spectrum at $5\ K$ reported by Ariu and co-workers (2003). The arrow indicates when F_{0v} becomes negligible.

Change in PFO molecular morphology reflects directly in the absorption spectrum. For amorphous PFO films with a fraction of the β-phase, the absorption spectrum contains an additional narrow band at 435 nm assigned by molecules in the β-phase as shown in Figure 6.7 (Ariu et al., 2003). From the synthesis of the oligo (9.9'-dioctylfluorenes) with different chain length ($n = 9 \sim 23$ monomers), there is an increase in the intensity of the β-phase absorption and its redshift occur as the value of n increases. It was found for β-phase formation for $n > 15$ monomers for a polymeric solution (Shiraki et al., 2015).

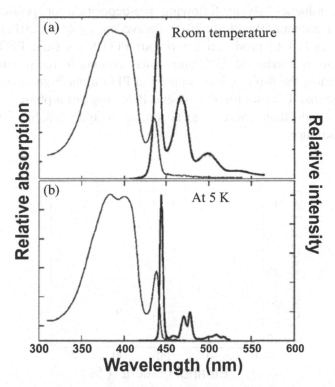

FIGURE 6.7 Absorption and photoluminescence spectrum for PFO film with 13% of β-phase generated by exposing amorphous films to toluene vapor. (a) Room temperature and (b) at 5 K. The sample was excited using *HeCd* laser at 354 nm in the photoluminescence measurement.

Source: Reprinted with permission from Ariu et al. (2003). Copyright (2003) by the American Physical Society.

A longer chain length promotes a lower energy gap than as described in Figure 6.2. Consequently, the PL spectrum redshift occurs (with peaks at 440, 468, 500 and 533 nm) due to the β-phase chains at room temperature. For example, emission at 440 nm which results from 0–0 transition is redshifted at 113 meV compared to the 0–0 transition for amorphous PFO (423 nm). Also, the β-phase chains exhibit a typical well-resolved PL line shape composed by seven emission peaks (444, 459, 471, 478, 501, 509, and 517 nm) at 5 K, as shown in Figure 6.7(b).

Although the β-phase can be directly induced using toluene vapor treatment, amorphous films are obtained by spin-coating technique from PFO in toluene solvent following pre-deposition or post-deposition thermal treatment (Bai et al., 2016; Perevedentsev et al., 2016). Bai and co-workers (2016) produced amorphous PFO films from PFO dilution in toluene by stirring at 75°C and slowly cooling to room temperature. In this study, the β-phase was induced in PFO films from ethanol mixed with precursor PFO solution in toluene. Percentage of β-phase in PFO film increases with ethanol percentage according to Figure 6.8, like findings for PFO in solution.

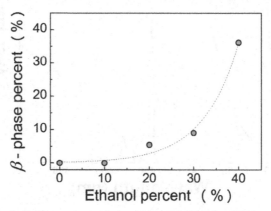

FIGURE 6.8 β-phase percentage induced in PFO film for different portions of ethanol in the precursor PFO solution (initial concentration 10 mg/ml in toluene solvent). For the preparation of the polymeric solution, the PFO was dissolved in toluene by stirring at 75°C and slowly cooled to room temperature.

Source: Adapted with permission from Bai et al. (2016). Copyright (2016) American Chemical Society.

It is also possible to control the β-phase percentage by mixing the paraffin oil in the precursor PFO solutions (Zhang et al., 2017). Also, PLQE shows a dependence on the β-phase fraction reaching maximum values around 69% for PFO films with 6% of the β-phase (Perevedentsev et al., 2016). Moreover, temperature strongly affects the PLQE of PFO films. At 10 K, amorphous PFO films show PLQE of 85%, while for PFO films containing 13% β-phase, the PLQE reduces to approximately 40% (Ariu et al., 2002). The energetic configuration of both phases promotes an efficient energy transfer between them affecting the exciton migration and consequently the PL.

6.4   EXCITON MIGRATION IN POLYFLUORENE

Energy transfer sequence of excited state between neighboring chromophores promotes the exciton migration in a conjugated polymer, also named as exciton transport. This process comprises one of the operating steps of organic photovoltaic cells (Clarke and Durrant, 2010). For example, in organic photovoltaic cells, photogenerated excitons are transported (during exciton lifetime) through the polymer up until polymer-charge acceptor interface, where the excitons are dissociated into charges.

A model introduced by Förster (1959) is commonly used to describe the energy transfer based on the dipole-dipole coupling. In this approach, the energy transfer rate (k_T) depends on the distance between the chromophores (R) by:

$$k_T = k_r (R_0/R)^6 \qquad (6.5)$$

where, k_r is the rate of the spontaneous radiative decay and R_0 is the radius of the energy transfer, also named as Förster radius. R_0 characterizes the distance at which the energy transfer rate is equal to the ratio of spontaneous radiative decay ($k_r \sim 10^9$ s^{-1} for conjugated polymer). Studies of the exciton dynamics using Monte Carlo simulation indicates that R_0 is approximately equal to 3 nm for disorder PFO (Meskers et al., 2001).

The behavior of the k_T against the distance between neighboring PFO chromophores is simulated for $R_0 = 3$ nm and is shown in Figure 6.9. Furthermore, for ranges less than 3 nm the energy transfer ratio assumes high values and decreases with R^6. When $R \sim 1$ nm, k_T becomes close to the non-radiative decay processes (k_{nr}).

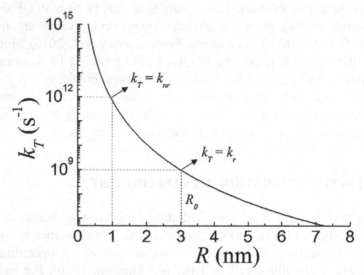

FIGURE 6.9   Rate of the energy transfer between neighboring polyfluorenes chromophores obtained from Eq. (6.5) using the typical radius of energy transfer for polyfluorenes (R_0 = 3 nm). The figure is plotted on a semi-logarithmic scale.

From photophysical process rates k_T, k_r, and the non-radiative rate k_{nr}, the probability (P_i) of any process (i) to occur (energy transfer, decay radiative and so on) at a time (t) is expressed by:

$$P_i = \frac{k_i}{k_{total}} \tag{6.6}$$

where, k_i is the specific process and k_{total} is the sum of all possible processes. In other words, for energy transfer and radiative decay, the probability is $P_T = k_T / (k_r + k_{nr} + k_T)$ and $P_r = k_r / (k_r + k_{nr} + k_T)$, respectively.

Considering a situation in which there are two single PFO chromophores with the same energy gap, separated by distance R and three photophysical processes with typical rates for conjugated polymer: (i) energy transfer (rate given by Eq. (6.5)), (ii) radiate decay ($k_r = 10^9$ s^{-1}) and (iii) the vibrational relaxation (VR) ($k_{nr} = 10^{12}$ s^{-1}). This approach provides a better understanding to analyze the probability of the photophysical processes in conjugated polymers. In this scenario, the probability of a possible photophysical process immediately after exciton photogeneration in one of the chromophores is shown in Figure 6.10 (top graph). The probability of VR (P_{nr}) increases with the increasing distance between chromophores. In contrast, the energy transfer probability decreases against greater values of R while the radiative decay is unlikely. At $R \sim 1$ nm, the energy transfer and the VR are the possible processes competing with probability approximately equal to 50%.

Bottom of Figure 6.10 shows the expected process when R is higher than R_0 and the probabilities as a function of R for further processes, assuming that a VR occurs. Energy transfer probability decreases while the radiative decay increases with the increase of R. The radius of the energy transfer R_0 = 3 nm is the critical distance between polyfluorene chromophores where the energy transfer and the radiative decay have the same probability equal to 50%, according to original interpretation by Förster (1959) for a general case. In other words, R_0 indicates the distance below which energy transfer is suitable to occur.

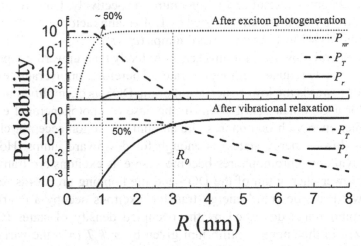

FIGURE 6.10 Probability of the photophysical processes (vibrational relaxation, energy transfer, and radiative decay), obtained from Eq. (6.6), plotted against the distance R between two single polyfluorenes chromophores with the same energy gap. P_{nr}, P_T, and P_r correspond to the probabilities of vibrational relaxation, energy transfer, and radiative decay, respectively. Top: photophysical processes probability immediately exciton generation. Bottom: the probability of photophysical processes after vibrational relaxation. The figure is plotted on a semi-logarithmic scale.

An ideal system to analyze some photophysical processes in PFO has been considered so far. However, the set of chromophores with different energy levels described by DOS should be considered for a better physical description. One consequence of the energetic disorder is the migration of the exciton from multiple energy transfers between the chromophores. Exciton migration can be interpreted via random hopping between chromophores (Laquai et al., 2009). In this model, exciton hopping rates are based on Miller-Abraham rates as follows:

$$W_{ij} = k^r \begin{cases} e^{-\left(\frac{\varepsilon_j - \varepsilon_i}{k_B T}\right)} & if \, \varepsilon_j > \epsilon_i \\ 1 & if \, \varepsilon_j \leq \epsilon_i \end{cases} \tag{6.7}$$

where, i and j indicate the acceptor and donor chromophores of the exciton, respectively, k_r is the energy transfer rate which depends on the distance R (given by Eq. (6.5)), ε is the chromophore energy, k_B, and T correspond to the Boltzmann constant and temperature, respectively. Upward jumps in energy ($\varepsilon_j > \varepsilon_i$) are thermally assisted by Boltzmann factor ($e^{-(\varepsilon j - \varepsilon i)/kBT}$) while this factor is set to 1 for downward jumps ($\varepsilon_j \leq \varepsilon_i$).

In general, downward jumps are more likely than upward jumps as the $e^{-(\varepsilon j - \varepsilon i)/kBT} < 1$, reduce the hopping rates. Therefore, on average, excitons proceed towards the lower energy tail of the DOS as exemplified in Figure 6.11 for a Gaussian distribution. In the case of photogenerated excitons in higher energy chromophores, the amount of exciton acceptor chromophores (lower energy states) is enough for downward jumps. However, these available chromophores become scarce as excitons are transported to the lower energy tail of the DOS and the hopping process is reduced. Afterward this downhill energy transfer, excitons occupy a distribution of chromophores described by the occupied density of states (ODOS) localized in the energy equilibrium given by-$\sigma^2/k_B T$ ($\sigma^2$is the variance of the DOS) (Laquai et al., 2009). All this migration process occurs during exciton lifetime, and the fluorescence comes from the chromophores contained in ODOS.

As a consequence of what has been described previously, a gradual redshift in time (*ps* scale) is observed in the fluorescence spectrum of the PFO under higher energy excitations in the absorption band (Meskers et al., 2001), as discussed below. This process is also named as spectral diffusion or relaxation of the excitation. Moreover, for excitation at 3.35 eV with maximum absorption around 3.3 eV (375 nm, as shown in Figure 6.5), there is a spectral diffusion of approximately 100 meV at 15 K for films of amorphous PFO.

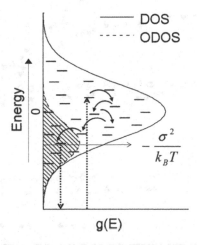

FIGURE 6.11 Illustration of the migration process in Gaussian DOS. Photogenerated excitons in higher energy chromophores undergo a downhill energy transfer for the chromophores in the DOS tail.

The effect of the energy excitation in the spectral diffusion is shown in Figure 6.12. In this experiment, the redshift was monitored via the energy of the 0–1 transition at 15 K as a function of time after excitation and for different excitation energies. While spectral diffusion is more evident for higher energy excitation, while there is a delay in spectral diffusion as the excitation energy decreases. For example, spectral diffusion occurs only above 300 ps under excitation at 2.591 eV. In this case, downward jumps prevail at low temperatures. However, unoccupied chromophores with lower energies are rare and on average, more distant.

Consequently, the delay for the downward jumps is expected because the exciton hopping rate depends on $1/R^6$. Photogenerated excitons in the lower energy states (excitation smaller than 2.92 eV) are trapped, and the migration process is limited.

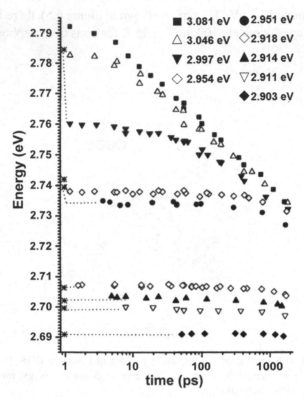

FIGURE 6.12   Spectral diffusion monitored by the energy of the 0–1 transition at 15 *K* of the polyfluorene amorphous films as a function of time after excitation.

Source: Reprinted with permission from Meskers et al. (2001). Copyright (2016) American Chemical Society.

Because of the migration process, the excitons travel an average distance during their lifetime. This average distance corresponds to the radius of the exciton migration (R_m), also known as exciton diffusion length. A typical path for *3D* exciton migration is shown in Figure 6.13 (red line) accompanied by a sphere, delimited by migration radius vector ($\vec{R}_m$). The value of $R_m = (9 \pm 2)$ nm was obtained for films derived from polyfluorene (poly(9,9-dioctylfluorene-co-N-(4-butylphenyl)-diphenylamine)) through ultrafast fluorescence decay measurements (Bruno et al., 2013). In general, amorphous polymers show values of exciton migration radius between 3.5 and 11 nm (Tamai et al., 2015).

The β-phase chains completely change the exciton migration in amorphous polyfluorene films. The low energy of the β-phase molecule can trap the excitons interrupting its migration process through the amorphous matrix, as illustrated in Figure 6.13 (exciton path indicated by black line). Ariu and co-workers (2003) proved a detailed description of the exciton migration in polyfluorene films containing β-phase, as discussed below. Even when the amorphous polyfluorene matrix is excited, the β-phase chains dominate the PL, an example can be found in Figure 6.7, and this can be attributed to the efficient energy transfer from the amorphous polyfluorene matrix to the β-phase chain. The authors suggest that more than 99% of the photogenerated excitons in the amorphous matrix are transferred to β-phase chains for polyfluorene films with 13% of the β-phase at 5 K.

Although the migration of excitons in the amorphous matrix can be interrupted, a spectral diffusion is observed from measures of time-dependent PL, suggesting the exciton migration through the β-phase molecules (Ariu et al., 2003). A transfer radius of 5.4 nm was estimated for energy transfer between β-phase molecules (Shaw et al., 2010) which is greater than amorphous polyfluorene films ($R_0 = 3$ nm). Therefore, it is expected that the energy transfer between β-phase molecules remains within the energy transfer radius of 5.4 nm. However, the narrow energy distribution of β-phase molecules implies a lower spectral diffusion. It was shown that spectral diffusion between β-phase chains is twice smaller than for amorphous polyfluorene film (Ariu et al., 2003).

6.5　CONCLUDING REMARKS

The spectroscopy and photophysical properties, with attention to the exciton migration, of the polyfluorene materials were described in this review chapter. From the interpretation based on energetic disorder produced by conformational disorder, the absorption and PL are interpreted. Amorphous PFO films show a broadband absorption spectrum in the blue region due to $S_0 \rightarrow S_1$ electronic transition accompanied by a vibrational shift of the set of chromophores contained in DOS, while the PL of amorphous PFO is characterized by 0–0 transition, followed by three vibronic transitions resulting from the Huang-Rhys parameter $S = 0.5$. The long chains of the β-phase molecular are generated in an amorphous polyfluorene matrix via different physicochemical treatments such as solvent vapor exposure, heating-cooling cycles, and poor solvent and non-solvent mixtures as discussed in this chapter. Polyfluorene films containing β-phase percentage show an additional absorption band at 435 nm and a well-defined PL spectrum at low temperatures, (seven emission peaks redshifted) both coming from β-phase chain. These features are the optical signature of the β-phase in polyfluorene.

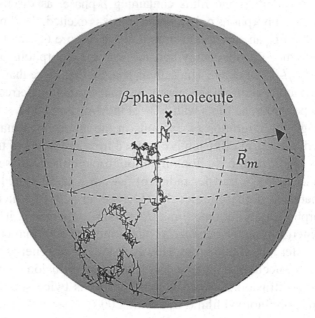

FIGURE 6.13   Illustration of the exciton migration in polyfluorene film with one β-phase site (indicated by xsymbol) within the sphere delimited by the average vector of migration ($\vec{R}_m$). Two excitons generated at the origin of the coordinate system and the red and black line correspond to the exciton path through an amorphous matrix without finding and finding β-phase site, respectively. The *3D* random walk is used to produce these expected paths of the excitons.

Energy transfer of the exciton in polyfluorene materials is successfully described using the Förtser model. The energy transfer radius of 3 nm characterizes the energy transfer process between polyfluorene chromophores in the amorphous matrix. While the radius of energy transfer between β-phase chromophores has a greater value of 5.4 nm. The multiple energy transfers between the polyfluorene chromophores promote the exciton migration. Excitons in PFO amorphous films show an average radius of movement of 9 nm. This process occurs by random hopping between chromophores supporting a downhill energy transfer for the energetic states of DOS tail. As a result, the experimental fluorescence spectrum is redshifted in time (*ps* scale). This effect is named as spectral diffusion. The spectral diffusion in amorphous polyfluorene films is greater for photogenerated excitons in higher energy chromophores. Alternatively, spectral diffusion is reduced when the excitons are photogenerated in lower energy chromophores. In polyfluorene films with 13% of *β*-phase chain, the spectral diffusion effect from *β*-phase PL is observed suggesting the exciton migration between *β*-phase chromophores.

KEYWORDS

- **density of state**
- **local exciton ground states**
- **occupied density of states**
- **photoluminescence quantum efficiency**
- **poly(9,9-dioctylfluorene)**
- **β-phase chromophores**

REFERENCES

Ariu, M., et al., (2002). The effect of morphology on the temperature-dependent photoluminescence. *Journal of Physics: Condensed Matter, 14*, 9975–9986.

Ariu, M., et al., (2003). Exciton migration in β-phase poly(9,9-dioctylfluorene). *Physical Review B., 67*(19), 1–11.

Bai, Z., et al., (2016). Quantitative study on β-phase heredity based on poly(9,9-dioctylfluorene) from solutions to films and the effect on hole mobility. *The Journal of Physical Chemistry C., 120*(49), 27820–27828.

Bansod, B. K., et al., (2017). A review on various electrochemical techniques for heavy metal ions detection with different sensing platforms. *Biosensors and Bioelectronics, 94*, 443–455.

Barford, W., & Marcus, M., (2014). Theory of optical transitions in conjugated polymers. I. Ideal systems. *Journal of Chemical Physics, 141*(16), 164101.

Barford, W., & Tozer, O. R., (2014). Theory of exciton transfer and diffusion in conjugated polymers. *Journal of Chemical Physics, 141*(16), 164103.

Barford, W., (2005). *Electronic and Optical Properties of Conjugated Polymers* (1st edn.). Nova York: Oxford University Press.

Beenken, W. J. D., & Pullerits, T., (2004). Spectroscopic units in conjugated polymers: A quantum chemically founded concept? *The Journal of Physical Chemistry B., 108*(20), 6164–6169.

Bruno, A., et al., (2013). Determining the exciton diffusion length in a polyfluorene from ultrafast fluorescence measurements of polymer/fullerene blend films. *Journal of Physical Chemistry C., 117*(39), 19832–19838.

Chen, S. H., Su, A. C., & Chen, S. A., (2005). Noncrystalline phases in poly (9,9-di-n-octyl-2,7-fluorene). *Journal of Physical Chemistry B., 109*(20), 10067–10072.

Choi, E. Y., et al., (2014). Photophysical, amplified spontaneous emission and charge transport properties of oligofluorene derivatives in thin films. *Physical Chemistry Chemical Physics, 16*(32), 16941.

Clarke, T. M., & Durrant, J. R., (2010). Charge photogeneration in organic solar cells. *Chemical Reviews, 110*(11), 6736–6767.

Forster, T., (1959). 10th Spiers memorial lecture. Transfer mechanisms of electronic excitation. *Discuss. Faraday Soc., 27*, 7–17.

Grell, M., et al., (1999). Interplay of physical structure and photophysics for a liquid crystalline polyfluorene. *Macromolecules, 32*(18), 5810–5817.

Guillén, M. G., et al., (2016). A fluorescence gas sensor based on forster resonance energy transfer between polyfluorene and bromocresol green assembled in thin films. *Sensors and Actuators, B: Chemical, 236*, 136–143.

Hu, D., et al., (2000). Collapse of stiff conjugated polymers with chemical defects into ordered, cylindrical conformations. *Nature, 405*(6790), 1030–1033.

Juang, R. S., et al., (2016). Enhanced sensing ability of fluorescent chemosensors with triphenylamine-functionalized conjugated polyfluorene. *Sensors and Actuators, B: Chemical, 231*, 399–411.

Knaapila, M., & Monkman, A. P., (2013). Methods for controlling structure and photophysical properties in polyfluorene solutions and gels. *Advanced Materials, 25*(8), 1090–1108.

Laquai, F., et al., (2009). Excitation energy transfer in organic materials: From fundamentals to optoelectronic devices. *Macromolecular Rapid Communications, 30*(14), 1203–1231.

Makhov, D. V., & Barford, W., (2010). Local exciton ground states in disordered polymers. *Physical Review B-Condensed Matter and Materials Physics, 81*, 1–6.

Marcus, M., Tozer, O. R., & Barford, W., (2014). Theory of optical transitions in conjugated polymers. II: Real systems. *Journal of Chemical Physics, 141*(16), 164102.

Meskers, S. C. J., et al., (2001). Dispersive relaxation dynamics of photoexcitations in a polyfluorene film involving energy transfer: Experiment and Monte Carlo simulations. *Journal of Physical Chemistry B., 105*, 9139–9149.

Nenashev, A. V., Oelerich, J. O., & Baranovskii, S. D., (2015). Theoretical tools for the description of charge transport in disordered organic semiconductors. *J. Phys.: Condens. Matter, 27*, 93201.

Oh, S. H., et al., (2010). Water-soluble polyfluorenes as an interfacial layer leading to cathode-independent high performance of organic solar cells. *Advanced Functional Materials, 20*(12), 1977–1983.

Peet, J., et al., (2008). Controlled β-phase formation in poly(9,9-di-n-octylfluorene) by processing with alkyl additives. *Advanced Materials, 20*(10), 1882–1885.

Perevedentsev, A., et al., (2016). Spectroscopic properties of poly(9,9-dioctylfluorene) thin films possessing varied fractions of β-phase chain segments: Enhanced photoluminescence efficiency via conformation structuring. *Journal of Polymer Science, Part B: Polymer Physics, 54*(19), 1995–2006.

Prins, P., et al., (2006). Enhanced charge-carrier mobility in β-phase polyfluorene. *Physical Review B-Condensed Matter and Materials Physics, 74*(11), 10–12.

Scholes, G. D., & Rumbles, G., (2006). Excitons in nanoscale systems. *Nature Materials, 5*, 683–696.

Shaw, P. E., et al., (2010). Exciton-exciton annihilation in mixed-phase polyfluorene films. *Advanced Functional Materials, 20*(1), 155–161.

Shiraki, T., et al., (2015). Strong main-chain length-dependence for the β-phase formation of oligofluorenes. *Polym. Chem., 6*(28), 5103–5109.

Tamai, Y., et al., (2015). Exciton diffusion in conjugated polymers: From fundamental understanding to improvement in photovoltaic conversion efficiency. *Journal of Physical Chemistry Letters, 6*(17), 3417–3428.

Yang, X., Dykstra, T.E., & Scholes, G.D. Photon-echo studies of collective absorption and dynamic localization of excitation in conjugated polymers and oligomers. *Phys. Rev. B., 71(4)*, 045203-1–045203-15

Yang, P. C., et al., (2016). Synthesis and chemosensory properties of two-arm truxene-functionalized conjugated polyfluorene containing terpyridine moiety. *RSC Adv., 6*(90), 87680–87689.

CHAPTER 7

Photophysical Properties of Poly(p-Phenylene Vinylene)

ROBERSON SARAIVA POLLI and RAQUEL APARECIDA DOMINGUES

Institute of Science and Technology, Federal University of São Paulo, 12231-280, São José dos Campos, SP, Brazil

7.1 INTRODUCTION

Common polymers as polystyrene (PS) and polyethylene (PE) have low conductivities, classified as insulators with σ bonds. The search of synthetic macromolecules with conductivities equal to or higher than metals has allowed the synthesis of dozens of chemically different macromolecules, each one with differing electrical properties. Normally, conducting polymers present alternated single and double (π) bonds, known as conjugated polymers. The π bands give the metallic or semi-metallic characters according to their completion. For instance, on polyacetylene, there are two sub-bands: the valence band π (occupied) and the conduction band π^* (not occupied) (Da Silva et al., 2008) (Figure 7.1).

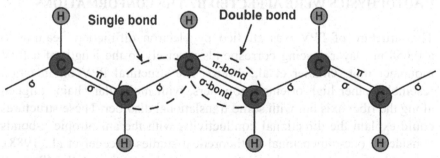

FIGURE 7.1 Conjugated double bonds.

Source: Reprinted from Le et al., 2017. Open access.

The polymer poly(p-phenylene vinylene) (PPV) is an organic semiconducting polymer with alternating phenylene and vinylene groups and therefore it is a conjugated polymer with high electrical conductivity (Masse, 2014). Besides its conductive properties, the PPV is a luminescent polymer with exciting electroluminescent properties, observed by Burroughes and coworkers (1990), with a bandgap of about 2.5 eV between their bands (π and π^*). The electrical, mechanical, and optical proprieties of this class of material were studied with a variety of experimental techniques, and these properties were related to the different morphologies of PPV (Masse, 2014).

Conjugated polymers usually have some well-known properties:

1. Absorption and emission range widely separated. The energy difference range between the absorption and emission region, the Stokes shift (SS), is usually wide in these polymers;
2. High efficiency in emission, with a high-speed relaxation process;
3. Broad photoluminescence (PL) peaks (excitons decays with phonons emissions) (Da Silva et al., 2008).

The conformations, temperature, and the way PPV is synthesized could affect the photophysical aspects, as absorption and emission bands, while theoretical models may explain some behaviors. Therefore, high molecular weight PPV films need to be investigated by several different methods, allowing PPV to be used in a series of applications.

7.2 PPV STRUCTURE, CONFORMATIONS AND HOW THE PHOTOPHYSICS WERE AFFECTED BY THE CONFORMATIONS

The structure of PPV was studied by electron diffraction, leading to a 0.658 nm layer spacing corresponds precisely to the length of a PPV monomer unit (Granier et al., 1986). The structural PPV organization resembles other high oriented polymers, with molecular chains aligned along the fiber axis but with some translational disorder. These structures could explain the directional conductivity, with the anisotropic π-bonds considered one-dimensional in theoretical studies (Heeger et al., 1988), but with transverse conductivities measured experimentally (Park et al., 1979). Although many properties of conducting polymers could be obtained with the one-dimensional conductivity, transverse conduction may refine the results.

Diffraction presents a monoclinic structure of PPV without dopants and orthorhombic structures with AsF_5, SbF_5, and H_2SO_4, indicating layers with polymer and dopant. The conducting crystals are not large, on the order of 5.0 to 7.0 nm equiaxed, for the undoped. The doped crystallites are also equiaxed and some of them smaller than 4.0 nm (Masse, 2014). The dopants agents may increase the electrical conductivity but, it could interfere with other properties such as optical and mechanical. PPV doped with AsF_5 and SbF_5, for example, has the tensile modulus and limit strain reduced (Machado, Masse, and Karasz, 1992).

Conjugated polymers as PPV have several optoelectronic applications, as a light-emitting layer in electroluminescence (EL) devices. Thus, it has been proposed new synthetic routes with the purpose to improve the PPV optical response. Halliday and coworkers obtained changes in the electronic structure, resulting in a strong redshift of the absorption, and blue-shifted of the PL spectra comparing to conventional PPV. These results were obtained using a soluble precursor polymer former under the controlled condition to produced hard road conjugated segments connected by flexible spacer groups (Halliday et al., 1996).

7.3   ABSORPTION AND EMISSION BANDS AND CONFORMATIONAL DEPENDENCE

As mentioned before, the forms and doping of PPV may change remarkably the optical properties. The PPV and an electrochemical doping form with a Li electrode immersed in an electrolyte solution of $LiBF_4$ were investigated by ESR, optical spectra, and electrical conductivity by Yoshino et al., (1986). The absorption spectra of the non-doped show a band-gab of 2.7 eV, intermediate between those of poly(p-phenylene) and polyacetylene. The linear dependence of $(\alpha h v)^2$ on $h v$ indicates a behavior of an insulator or semiconductor with a direct band-gap. With the BF_4^- the absorption spectrum changes, suppressing the inter-band transition and creating new absorption peaks around 0.7 and 2.0 eV, shifted to higher energy side according to the increase in the dopant concentration (Yoshino et al., 1986).

In other work, Steger-Smith studied PPV films prepared with two different sulfonium salt precursors, one based on a cyclic and acyclicsulfide, by fluorescence, Raman, and X-ray scattering (Stenger-Smith et al., 1989). The fluorescence in UV/Vis of the PPV with the cyclic sulfide, unlike the acyclic one, was different from the previously reported spectra for PPV, presenting phonon sidebands at 20,220, 21,740, and 23,000 ± 100 cm^{-1}. However, the experimental gap of 2.4 eV agrees with theoretical values of 2.5 eV (Brédas et al., 1982). The phonons sidebands of the PPV with the cyclic precursor were also obtained at room temperature, although more sharply at liquid nitrogen temperatures. The Raman spectrum was impaired by a large amount of fluorescence, allowing only two distinct peaks (at 1172 and 1584 cm^{-1}), identical for PPV made from both precursors. The first peak is related to centrosymmetric C-H in-plane bend and the last one, to the double-bond stretch.

The photoexcited states in the conjugated polymer of PPV, poly (2, 5-dimethoxy-p-phenylene vinylene) also known as PDMeOPV were studied by Woo and coworkers (1992). In this work, the authors compared a film of PDMeOPV with a spin-coated PPV sample (prepared with the tetrahydrothiophenium precursor). At room temperature, the π-π^* absorption in the first example is observed at 2.0 eV, around 15% lower than for the PPV sample. At 80 K, there were obtained three peaks photoinduced absorptions (PAs) at 0.68, 1.35, and 1.8 eV. The first and last were long-lived (with significant numbers surviving more than 100 ms) and assigned to charged bipolarons, and the other one has a lifetime of order 2.5 ms and maps to a triplet-triplet transition.

PPV and its derivatives exhibit strong luminescence effect showing a strong dependence on intrinsic defects, on molecular aggregations and their structural order. Marlleta and coworkers (2002) analyze emission line shapes, measured at 30 K and at room temperature, of self-assembly (SA) films of PPV using a semi-empirical model, which assumes that defects along the molecule give rise to a distribution of conjugated segments of different lengths. The PL spectrum of a SA-PPV film in 500–625 range shows a peak around 521 nm fitted by the zero-phonon transition and other 3 unresolved structures at 535, 555 and 570 nm, corresponding to vibrational modes, respectively to 500 cm^{-1} (62 meV), 1170 cm^{-1} (139 meV) and 1550 cm^{-1} (192 meV). This model assumes that defects along the molecule give rise to a distribution of conjugated segments of different lengths. There was a slight difference in the theoretical model and the experimental data, indicating that other interactions need to be considered (Marletta et al., 2002).

A complete study of the optical proprieties of PPV and their relation-ship of the degree of intrachain order were performed by Pinchler and coworkers (1993). They used an improved form of PPV (more ordered) and two types of standard PPV. There were slight differences between the usual forms, thus were compared the improved form with just one of the standard samples, named Type II (the one with better-ordered structure). The optical absorption between 1.0 and 3.5 eV shows the $S_0 \rightarrow S_1/0 \rightarrow 0$ transition at transition is at around 2.55 eV, and the maximum of absorp-tion is at 2.85 eV, while in the improved PPV, the peak is redshifted with $S_0 \rightarrow S_1$ transition at 2.463 eV, both at room temperature. The increased order in the improved PPV, as expected, change the character and distribution in energy of the singlet exciton. The authors attributed the difference to the conjugation length distribution: heavily weighted in the direction of long conjugation lengths in the improved PPV and short conjugation lengths in the standard PPV. Luminescence spectra for both samples were similar with emission peaks, and their vibronic progression in the two forms of PPV is almost identical at room temperature and 15 K, with the separation between $S_0 \rightarrow S_1/0 \rightarrow 0$ and $S_0 \rightarrow S_1/0 \rightarrow 1$ of around 180 meV. Unlike the emission spectra, that involve migration of the excited states to the lowest energy, in this case, the most conjugated, the absorption spectra show the contribution from all polymer chain segments, sampling the distribution of conjugation lengths. Thus, only the absorption spectra showed differences (Pichler and Halliday, 1993).

The PA was obtained in a range of temperatures, pump intensities, and chop frequencies for both samples, detecting only one peak at 1.36 eV and 1.38 eV for the improved and the standard PPV, respectively. This peak was assigned to a triplet-triplet absorption, slightly redshifted in comparison with standard PPV on the literature, but in agreement with the redshifts of other optical transitions in the same work. At 0.6 and 1.6 eV, transitions have been observed in different samples of PPV associated with photogenerated bipolarons, but not on the improved and this standard samples. The authors assigned this behavior due to a combination of the generation rate of the excitations and decay kinetics, mainly the reduction of generation rate of these states (Pichler and Halliday, 1993).

Da Silva and coworkers also studied PPV derivative thin film poly[2-methoxy-5-(20-ethyl-hexyloxy)-1,4-phenylene vinylene] (MEH-PPV), produced by a spin-coating technique at rotational speeds of 300, 1000, and 4000 rpm (A300, A1000, and A4000 samples) (Da Silva et al., 2008).

All samples present two high energy peaks and a shoulder at low energy region. The peak at 2.12 eV (0–0) is assigned to an electronic transition, while the peak at 1.98 eV and the shoulder at 1.78 eV assigned to vibronic bands, (0–1) and (0–2) respectively. The A300 electronic transition peak was slightly shifted (a decrease of around 30 meV) compared to the A4000, probably due to the smaller thickness and high homogeneity of the A4000 sample (Da Silva et al., 2008).

The absorption and emission spectra in conjugated polymers are strongly weighted by their strong coupling between molecular electronic states and vibrational states and uncertainty related to molecules conjugation length. In a simple way, the optical transitions in conjugated polymers depend on same parameters as: the position, intensity, and full width at half maximum (FWHM) of the electronic change $(0 \rightarrow 0)$ and the energy and Huang-Rhys (S_i) factor of each vibrational mode (Da Silva et al., 2008). While the positions, intensity, and the Huang-Rhys factors are related to energy and transitions (electronic and vibrational), the FWHM assigned the dispersion of conjugation length of the absolvers and emission chains.

7.4 TEMPERATURE EFFECTS

The temperature could alter the optical proprieties significantly and help to understand the structure and dynamics of bonds and interactions on the molecule. The temperature dependence of PL spectra of two polymer light-emitting diodes PPV derivatives based on poly[2,5-bis(3,'7'-dimethyl-octyloxy)1,4-phenylene-vinylene] (BDMO-PPV). And poly[2-methoxy-5(2'-ethyl-hexoxy)1,4-phenylene-vinylene] (MEH-PPV) were studied by Wantz and coworkers (2005). In both samples, the FWHM increased, with increasing temperature while the zero-phonon line was blue-shifted. The blue shift phenomenon correlates with the reduction of effective conjugation length, mainly with the increase of the Huang-Rhys parameter. However, this was not the case for the MEH-PPV sample, due to its almost temperature-independent Huang-Rhys parameter. The authors proposed a model of thermally activated statistic occupation of excited states, with a good approximation with the experimental data (Wantz et al., 2005) (Figure 7.2).

Snedden and coworkers (2010) also study the temperature dependence in a derivative of PPV, the PPV copolymer super yellow (SW), widely used in electro-optical devices. The temperature-dependent steady-state luminescence allowed estimating average conjugation length and the conformational energy. The decrease in temperature, starting at room temperature, showed redshifted PL and enhancement of the pure electronic transition peak comparing to the first vibronic peak. This improvement is related to the conjugation length, as the weakening in electron-vibrational mode interaction at low temperatures corresponds to a high structural order due to the electron delocalization. In this work, the FWHM of the electronic transition is practically independent of temperature, unlike the reduction observed in other PPV derivatives (Wantz et al., 2005).

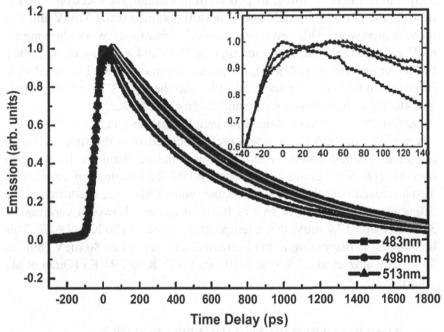

FIGURE 7.2 Photoluminescence decay for a spin-coated solid thin film of the PPV copolymer super yellow (SW).

Source: Reprinted with permission from Snedden, 2010. © Elsevier.

The internal conversion (IC) dynamics in conjugated polymers also could be affected by the temperature. In this recent work (Zhang et al., 2018), the PPV was used as the molecule model system, modeled as a one-dimensional chain. In this work, the blue shift with the increasing temperature was not observed. The model was limited, considering a single polymer chain and not including the thermal-induced interchain packing disorder, not allowing contribution of conjugation along the PPV chain, assigning its importance to explain the blue shift behavior. However, following stimulated absorptions, the IC from a higher-lying excited state to the lowest-lying exciton state showed a strong correlation with the temperature.

In Da Silva work, at low temperatures, the PL spectra of sample A 1000 (1000 rpm) show a narrow peak for the electronic transition and a series of vibronic sidebands which reveal the electron coupling with two different vibronic modes (Da Silva et al., 2008). The data were obtained in the temperature range of 130 K to 290 K. In this range, the electronic transition and the first vibrational band blue shift obtained was 86 meV and 105 meV, respectively. This result was coherent with Quan work showing 60 meV and 120 meV, but with the range of 77 to 294 K (Quan et al., 2006). In Pichler and Halliday (1993), the peak absorption of the improved PPV is redshifted from that for standard PPV, with the $S_0 \rightarrow S_1/0 \rightarrow 0$ transitions at 2.463 eV at room temperature and 2.416 eV at 80 K.

Da Silva work also obtained an increase of the electronic line intensity with the decrease in temperature, associated with the suppression of radiative channels and activation of non-radiative channels, likely multi phonons (Da Silva et al., 2008). The FWHM for the electronic transition also increased with the temperature, but with a difference according to the fitting method. The use of free or fixed Gaussians showed a variation of 34.8 meV and 17.7 meV, in the temperature range of 130 K to 290 K. This broadening is dispersing in the literature, with results as 80 meV (80K to 350 K) (Wantz et al., 2005) and 105 meV (77 K to 294 K) (Quan et al., 2006).

7.5 TIME RESOLVED PHOTOLUMINESCENCE (PL)

The PL decay depends on factors as carrier dynamics, oxidation processes, and quenching mechanisms. The quenching mechanism refers to the attribution of the additional non-radiative channel to excitons, assign to the creation of carbonyl groups (Yan et al., 1994). However, it is possible to achieve an enhancement of PL in PPV. Gobato and coworkers (2013) used a PPV film (1.1 μm) exposed to a laser (458 nm and 254 mW/cm^2) in the air and obtained an enhancement of 200% of PL intensity after 46 min of exposure. The authors assign this to carbonyl incorporation, shortening the chain allowing efficient spectral diffusion of excited carriers to a non-degraded PPV layer.

The same findings were obtained by Anni and coworkers (2003) but they also got in the transient PL measurements a bi-exponential decay: one with fast decay time (around 50 ps) assigned to diffusion on defects and a slower decay time (varies with the sample, but at least more than three times the fast decay time) attributed to radiative recombination.

The way the thin film is obtained and its thickness can change its absorbance and luminescence spectra. Laureto and coworkers (2012) analyzed PPV films produced by the layer-by-layer method. Ultra-thin films, with 2 to 3 layers, presents both absorbance and luminescence spectra shifted to high energy region, compared to thicker films. However, if obtained a thick film by adding layers to ultra-thin films, the shifted is maintained. Thus, adding layers could modulate the film energy profile (Laureto et al., 2012). In other work with PPV obtained by the layer-by-layer method (Marletta et al., 2003), Marlleta, and coworkers observed the absorbance and PL of a thin film PPV obtained with 20 layers. The film presented high thermal stability, a small blue shift of 5 nm in the zero-phonon peak and low electron-phonon coupling. In other work (Marletta and Gonçalves, 2006), Marletta, and coworkers studied the absorption and emission spectra of PPV-dodecylbenzenesulfonic counter-ion (DBS) films obtained by spin coating. According to the amount of DBS used in the preparation of the films, a shift of the center of the segment distribution to more considerable conjugation lengths was observed. Thus, the amount of DBS could control the conjugation degree of PPV.

The PL decay and the PL excitation spectra of polymers as PPV could change under UV-photodegradation. Kock and coworkers (1999) measured the UV-photodegradation in PPV and PPV derivatives poly-4,4'-diphenylene diphenylvinylene (PDPV) and poly-(1,3-phenylene diphenylvinylene) m-PPV-DP in the presence of oxygen. PPV PL showed remarkable differences when compared the UV-photodegradation take place in air or a vacuum. Unlike in the vacuum, with an almost exponential decay, in the air, the PL had an initial rapid decline, followed by a rise after about 20 s before decaying over again, practically linearly with irradiation time. During the firsts 120 s, the PL peak shifts hypsochromically from 2.228 to 2.234 eV. The hypsochromic shift could be associated with a reduced conjugation length with irradiation time; however, as this shortening in conjugation length could not explain the rise of the PL peak, the authors suggest that the steric interaction between the oxygen and the polymer may explain this behavior (Koch et al., 1999) (Figure 7.3).

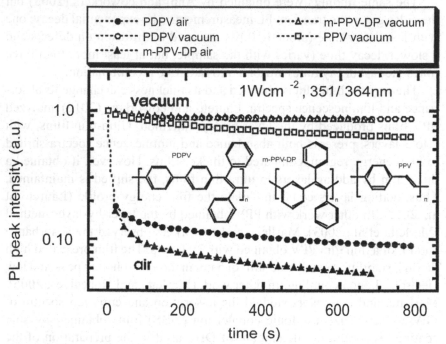

FIGURE 7.3   UV-photodegradation in PPV and PPV derivatives in air and vacuum.

Source: Reprinted from permission from Koch et al., 1999. © Elsevier.

7.6   THEORETICAL MODELS

The use of theoretical models helps the understanding of the molecule structure and dynamics. Zhang and coworkers study showed how the IC dynamics were affected by temperature in a conjugated polymer, with the PPV been used as the molecule model. The theoretical model was a modified version of the Su-Schrieffer-Heeger model (SSH) (Su et al., 1979) that used the simplest conjugated polymer molecule polyacetylene. In this earlier study, data as the energy of formation and activation energy for the motion was obtained, giving some light in the charge-transfer (CT) mechanism. Zhang used the same model, however, including the temperature effect. The PPV was modeled as a one-dimensional chain, with the Hamiltonian separated in annihilation and creation operators and kinetic and elastic energy, while the temperature was controlled by a laser, modeled as a pump pulse. Using a 40-site PPV chain, they obtained the energy gap between the lowest unoccupied molecular orbital (LUMO) and the highest occupied (HOMO) of 2.8 eV, like others results in the literature (Mizes, 1994).

Mizes (1994) was the first one to study the photoinduced charge transfer in PPV and used the same SSH method. This Hamiltonian is suitable to describe polarons, due to the lack of Coulomb interaction explicit. They suggest that, above the absorption edge, there are formed polarons or excitons, related to the picosecond absorption. This property is found even in parallel PPV chains with two monomers in length, although the length could change the number of pairs and the PA (Mizes, 1994).

Theoretical models were also applied in the study of stretched PPV films. In their work, Ramos, and coworkers experimentally found that the PL depends on the direction of the excitation beam of polarized light, perpendicular or parallel to the stretched axis (Ramos et al., 2015). The theoretical data were obtained from a Classical Molecular Dynamics approach, the Universal Force Field (Rappé et al., 1992), extracting features as linearity, local alignment, and ordering of neighbor chain segments. The local packing was also studied through a quantum semi-empirical method (Ramos et al., 2015). This work shows that the choice of the theoretical model depends on the property studied and not only the polymer structure.

Photoisomerization was also studied in vinylenes by theoretical models, as Liu and coworkers' study (Liu et al., 2011). In this study, the charge transport of cis and trans isomers of 2,5-diphenyl-1,4-distyrylbenzene (DPDSB) was analyzed. Depends on the measurements proprieties as temperature, there are more suitable models than others. In the charge transport case, there are two kinds of methods: a band like and hopping. The last one is more suitable for measurements with a high disorder, as in the room temperature (Coropceanu et al., 2007). The charge hopping methods could also be classified in two kinds of methods: Miller-Abrahams and Marcus, where the last one is valid for large electron-phonon (vibration) couplings and high temperature (Coropceanu et al., 2007). Thus, in Liu and coworkers' study, was utilized the semi-classical Marcus hopping model. The transition of cis and trans isomers showed a remarkable difference between electron and hole transport, almost balanced in the trans isomer but with a considerable hole transport compared to electron transport in the cis isomer (Liu et al., 2011).

7.7 PPV APPLICATIONS

Conjugated polymers as PPV are usually applied in semiconducting devices and bioimaging probes. Some of these applications will be shown in this chapter, demonstrating the high versatility of this molecule.

PPV was the first conjugated polymer in which EL was studied in detail, due to its ease of fabrication and excellent structural properties. The light emitted in this case is green-yellow and the gap between bonding (π) and antibonding (π^*) states is below2.5 eV (Burroughes et al., 1990). Besides the low-cost manufacturing of films, the well-known electronic properties and mechanisms in polymers are an advantage compared to inorganic semiconductors. In an organic light-emitting diode, the indium (ITO) is used as a transparent electrode, allowing the light to leave the device; while the top electrode is obtained by thermal evaporation of a metal (Burroughes et al., 1990). This kind of invention uses a hole injecting electrode and an electron injecting electrode. ITO is used as a hole injecting electrode due to its high work function and metals with low work function as Mg, and Al are used as an electron injection electrode (Friend et al., 1999). An application of organics LEDs is the use in displays, especially in the full-color graphics display. Problems as colors chromaticity, new addressing schemes, and methods of color patterning in these LEDs have been addressed, what could make organic LEDs more commercially successful than the liquid crystal ones.

The electric properties of conjugated polymers as PPV also allowed their use as organic light emitting transistor (OLET). The OLET has the feature of switching the electronic signal as a field-effect transistor (FET) and generates light, whose color depends on the organic material used in its fabrication (Reshak et al., 2013). The MEH-PPV was used in OLEDs and OLETs, mainly due to its low-cost processing and being easy to dilute in volatile aromatic and non-aromatic solvents. Reshakand coworkers showed that the spin rotation deposition led to differences in current-voltage characteristics, important in the optimization of the device (Reshak et al., 2013). Poly[2-methoxy-5-(2'-ethylhexyloxy)-p-phenylene vinylene) (MEH-PPV) thin films were also used in field configurable transistor (FCT), a kind of configurable FET, in a silicon nanowire FET platform. This type of transistor, have high plasticity and flexible configurability, facilitating field-programmable circuits, including a synaptic electronic circuit with dynamic learning functions (Lai et al., 2008).

PPV can also be applied in photovoltaic devices, mainly solar cells (Alem et al., 2004; Jenekhe and Yi, 2000; Sariciftci et al., 1993). Janeke and Yi have made solar cells from heterojunctions of PPV and poly(be nzimidazobenzophenanthroline ladder) (BBL). In these solar cells, the polymers layer thickness changes the charge collection efficiency, with the maximum obtained at 50 nm, for both PPV and BBL layers, with a five times reduction when the BBL changes to 75 nm, demonstrating the importance of bilayer thickness as a variable in their optimization (Jenekhe and Yi, 2000). In other work, Alem, and coworkers showed that in solar cells, with MEH-PPV and buckminsterfullerene C_{60}, the surface treatment in an interpenetrated network has the potential for improving efficiency (Alem et al., 2004).

PPV was also used in other areas, especially with its fluorescence characteristic associated with nanoparticles (NPs). Fluorescents PPV NPs were used as a fluorescence staining to enhance the contrast of fingerprint images obtained through the fuming process. The most used fluorescent reagent is Rhodamine 6G and its derivatives, but this kind of reagent emits red fluorescence and absorbs far from the usual portable UV lamps, while PPV can be excited by the UV lamp (Chen et al., 2018).

Besides the optical properties, conjugated polymers also show biocompatibility, allowing their use in bioimaging (Peters et al., 2018, 2016). PPV NPs presented high photostability and fluorescence brightness, and it was used in cells of the central nervous system, showing biocompatibility and surface charge associated with the cellular uptake of NPs (Peters et al., 2016). However, as expected, factors as size and the material choice could decrease the cell uptake, with a limit of around 20 nm (Peters et al., 2018).

7.8 CONCLUDING REMARKS

The spectroscopy and photophysical properties of the PPV and PPV derivatives materials were described in this review chapter. Aspects as structure, conformations, temperature effects, theoretical models, and applications are interpreted.

The conformations, temperature, and the way PPV is synthesized could affect the photophysical aspects, as absorption and emission bands. Naturally, the optical transitions in conjugated polymers depend on parameters like the position, intensity, and FWHM of the electronic transition ($0 \rightarrow 0$) and the energy and Huang-Rhys (S_i) factor of each vibrational mode. The use of theoretical models helps the understanding of the molecule structure and dynamics. Using a 40-site PPV chain, they obtained the energy gap between the LUMO and the highest occupied (HOMO) of 2.8 eV.

Conjugated polymers as PPV have several optoelectronic applications, in particular as a light-emitting layer in EL devices. Thus, it has been proposed new synthetic routes with the purpose to improve the PPV optical response.

KEYWORDS

- **field configurable transistor**
- **field-effect transistor**
- **organic light-emitting diodes**
- **photophysics**
- **poly(p-phenylene vinylene)**
- **polymer**

REFERENCES

Alem, S., et al., (2004). Efficient polymer-based interpenetrated network photovoltaic cells. *Applied Physics Letters, 84*(12), 2178–2180.

Anni, M., et al., (2003). Defect-assisted photoluminescence intensity enhancement in poly(p-phenylene vinylene) films probed by time-resolved photoluminescence. *Physical Review B-Condensed Matter and Materials Physics, 68*(11), 3–8.

Brédas, J. L., et al., (1982). Ab initio effective Hamiltonian study of the electronic properties of conjugated polymers. *Journal of Chemical Physics, 76*(7), 3673–3678.

Burroughes, J. H., et al., (1990). Light-emitting diodes based on conjugated polymers. *Nature, 347*(6293), 539–541.

Chen, H., et al., (2018). Fluorescence development of fingerprints by combining conjugated polymer nanoparticles with cyanoacrylate fuming. *Journal of Colloid and Interface Science, 528,* 200–207.

Coropceanu, V., et al., (2007). Charge transport in organic semiconductors. *Chemical Reviews, 107*(4), 926–952.

Da Silva, M. A. T., et al., (2008). Optical properties of thin films of Meh-PPV produced by the spin-coating technique at different rotational speeds. *Semina Ciências Exatas e Tecnológicas,* [s. l.], *29,* 15–38.

Friend, R. H., et al., (1999). Electroluminescence in conjugated polymers. *Nature, 397*(6715), 121–128.

Gobato, Y. G., et al., (2013). Photo induced photoluminescence intensity enhancement in poly(p-phenylene vinylene) films. *Applied Physics Letters, 942*(2002), 1–4.

Granier, T., et al., (1986). Structure investigation of poly(p-Phenylene Vinylene). *Journal of Polymer Science Part B: Polymer Physics, 24,* 2793–2804.

Halliday, D. A., et al., (1996). Large changes in optical response through chemical pre-ordering of poly(p-phenylenevinylene). *Advanced Materials,5*(1), 657–659.

Heeger, A. J., et al., (1988). Solitons in conducting polymers. *Reviews of Modern Physics, 60*(3), 741–850.

Jenekhe, S. A., &Shujian, Y. I., (2000). Efficient photovoltaic cells from semiconducting polymer heterojunctions. *Applied Physics Letters, 77*(17), 2635–2637.

Koch, A. T. H., et al., (1999). Enhanced photostability of poly (1, 3-phenylene diphenylvinylene) derivatives by diphenyl-substitution. *Synthetic Metals, 100,* 113–122.

Lai, Q., et al., (2008). An organic/Si nanowire hybrid field configurable transistor. *Nano Letters, 8*(3), 876–880.

Laureto, E., et al., (2012). Thickness effects on the optical properties of layer-by-layer poly(p-phenylene vinylene) thin films and their use in energy-modulated structures. *Current Applied Physics, 12*(3), 870–874.

Le, T.-H.; Kim, Y.; Yoon, H. Electrical and Electrochemical Properties of Conducting Polymers. Polymers 2017, 9, 150. https://www.mdpi.com/2073-4360/9/4/150

Liu, D., et al., (2011). Cis-and trans-isomerization-induced transition of charge transport property in PPV oligomers. *Chemical Physics, 388*(1–3), 69–77.

Machado, J. M., Masse, M. A., &Karasz, F. E., (1992). Anisotropic mechanical properties of uniaxially oriented electrically conducting poly(p-phenylene vinylene). *Polymer, 30,* 1992–1996.

Marletta, A., & Gonçalves, D., (2006). Analysis of the absorption and emission spectra of poly(p-phenylene vinylene) films thermally converted at a relatively low temperature. *Journal of Non-Crystalline Solids, 352*(32–35), 3484–3487.

Marletta, A., et al., (2003). Enhanced optical and electrical properties of layer-by-layer luminescent films. *Journal of Applied Physics,94*(9), 5592–5598.

Marletta, A., Guimarães, F. E. G., &Faria, R. M., (2002). Line shape of emission spectra of the luminescent polymer poly(p-Phenylene Vinylene). *Brazilian Journal of Physics,32*(2B), 570–574.

Masse, M. A., (2014). *Structure and Morphology of Electrically Conducting Poly(p-Phenylene Vinylene).* Doctoral Dissertation, University of Massachusetts Amherst, Amherst, USA. Retrieved From https://scholarworks.umass.edu/cgi/viewcontent. cgi?referer=https://www.google.com/&httpsredir=1&article=1759&context=dissertat ions_1 (accessed on 17 June 2020).

Mizes, H. A. M., & Conwell, E., (1994). Photoinduced charge transfer in poly(p-phenylene vinylene). *Physical Review B.* [s. l.], *50*(15), 243–246.

Park, W. Y., et al., (1979). Anisotropic electrical conductivity of partially oriented polyacetylene. *Journal of Polymer Science: Polymer Letters Edition, 17*(5), 195–201.

Peters, M., et al., (2016). PPV based conjugated polymer nanoparticles as a versatile bioimaging probe : A closer look at the inherent optical properties and nanoparticle-cell interactions. *Biomacromolecules, 17*(8), 2562–2571.

Peters, M., et al., (2018). Size-dependent properties of functional PPV-based conjugated polymer nanoparticles for bioimaging. *Colloids and Surfaces B: Biointerfaces, 169*, 494–501.

Pichler, K. H. D., (1993). Optical spectroscopy of highly ordered poly(p-phenylene vinylene). *Journal of Physics: Condensed Matter, 5*, 7155–7172.

Ramos, R., et al., (2015). Polarized emission from stretched PPV films viewed at the molecular level. *Physical Chemistry Chemical Physics, 17*(32), 20530–20536.

Rappé, A. K., et al., (1992). UFF, a full periodic table force field for molecular mechanics and molecular dynamics simulations. *Journal of the American Chemical Society, 2*(114), 10024–10035.

Reshak, A. H., et al., (2013). Electrical behavior of MEH-PPV based diode and transistor. *Progress in Biophysics and Molecular Biology, 113*(2), 289–294.

Sariciftci, N. S., et al., (1993). Semiconducting polymer buckminsterfullerene heterojunctions: Diodes, photodiodes, and photovoltaic cells. *Applied Physics Letters, 62*(6), 585–587.

Snedden, E. W., et al., (2010). High photoluminescence quantum yield due to intramolecular energy transfer in the super yellow conjugated copolymer. *Chemical Physics Letters, 490*(1–3), 76–79.

Stenger-Smith, J. D., Lenz, R. W., & Wegner, G., (1989). Spectroscopic and cyclic voltammetric studies of poly(p-phenylene vinylene) prepared from two different sulfonium salt precursor polymers. *Polymer, 30*(6), 1048–1053.

Su, W. P., Schrieffer, J. R., &Heeger, A. J., (1979). Solitons in polyacetilene. *Physical Review Letters, 42*(25), 1698–1701.

Wantz, G., et al., (2005). Temperature-dependent electroluminescence spectra of poly(phenylene-vinylene) derivatives-based polymer light-emitting diodes. *Journal of Applied Physics, 034505*(2005), 1–6.

Yan, M., et al., (1994). Defect quenching of conjugated polymer luminescence. *Physical Review Letters, 73*(5), 744–747.

Yoshino, K., et al., (1986). Electrical and optical properties of poly(p-phenylene vinylene) and effects of electrochemical doping. *Japanese Journal of Applied Physics, 25*(6), 881–884.

Zhang, M., et al., (2018). Temperature effect on the internal conversion dynamics following different stimulated absorptions in a conjugated polymer. *Organic Electronics: Physics, Materials, Applications, 56*, 201–207.

Index

Printed in the United States
by Baker & Taylor Publisher Services